Biopolitics and Utopia

Palgrave Series in Bioethics and Public Policy

Biotechnology continues to impact populations in myriad ways—influencing contemporary issues in food supply, genetic therapy, health care, biosecurity, terrorism, criminal justice, food supply, and environmental engineering, among many other aspects of daily life. The Palgrave Series in Bioethics and Public Policy seeks to promote interdisciplinary research that analyzes and assesses the social, environmental, and moral ramifications of where this technology is taking us. With a wide range of topics within bioethics open to the series, this series provides a home for cutting-edge research that bridges the divide between the natural and social sciences. This series will also attract a dynamic and varied assortment of scholars to provide comprehensive evaluations of where biotechnology is taking our society—and most importantly, if these directions are being forged appropriately and ethically.

Series editor: Sheldon Krimsky

Sheldon Krimsky is the Lenore Stern Professor of Humanities and Social Sciences and adjunct professor of Public Health and Community Medicine at Tufts University, USA. Professor Krimsky is the author, coauthor, and editor of 14 books including *Genetic Justice: DNA Databanks, Criminal Investigations, and Civil Liberties*, awarded a gold medal by the Independent Publishers in 2011. Professor Krimsky served on the National Institutes of Health's Recombinant DNA Advisory Committee and was a consultant to the Presidential Commission for the Study of Ethical Problems in Medicine and Biomedical and Behavioral Research and to the Congressional Office of Technology Assessment. Recently, he served as associate editor, for *Bioethics, 2014*, a reference volume for the field.

Regulating Preimplantation Genetic Diagnosis in the United States:
The Limits of Unlimited Selection (2015)
   By Michele Bayefsky and Bruce Jennings

Biopolitics and Utopia: An Interdisciplinary Reader (2015)
   Edited by Patricia Stapleton and Andrew Byers

Biopolitics and Utopia
An Interdisciplinary Reader

Edited by
Patricia Stapleton and Andrew Byers

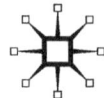

BIOPOLITICS AND UTOPIA
Copyright © Patricia Stapleton and Andrew Byers, 2015.
All rights reserved.

First published in 2015 by
PALGRAVE MACMILLAN®
in the United States—a division of St. Martin's Press LLC,
175 Fifth Avenue, New York, NY 10010.

Where this book is distributed in the UK, Europe and the rest of the world, this is by Palgrave Macmillan, a division of Macmillan Publishers Limited, registered in England, company number 785998, of Houndmills, Basingstoke, Hampshire RG21 6XS.

Palgrave Macmillan is the global academic imprint of the above companies and has companies and representatives throughout the world.

Palgrave® and Macmillan® are registered trademarks in the United States, the United Kingdom, Europe and other countries.

ISBN: 978–1–137–51474–5

Library of Congress Cataloging-in-Publication Data

   Biopolitics and utopia : an interdisciplinary reader / edited by Patricia Stapleton and Andrew Byers.
      pages cm.—(Palgrave series in bioethics and public policy)
   Summary: "Biopolitics and Utopia explores the intersection of biopolitics and utopian thought. As an interdisciplinary work, it addresses many salient biopolitical issues (state and medical interventions in the body, fears over scientific progress, resistance to state biopower, and ethical concerns), while also engaging in the utopian drive behind biopolitical efforts. The book is structured into four main sections: Actions, Speculations, Reactions, and Reflections. The chapters in Actions examine the practices of direct, medical intervention to 'normalize' citizens' bodies. The next section, Speculations, approaches the intersection of utopia and biopolitics through a literary lens, reviewing science fiction texts as expressions of cultural and social fears about scientific progress. Reactions outlines potential acts of resistance in the face of biopower. Finally, Reflections offers a more philosophical essay, which engages the reader in the potential for creating an ethics for scientific standards"—Provided by publisher.
   Summary: "This reader offers a fascinating exploration of the intersection of biopolitics and utopia by employing a range of theoretical approaches. Each essay provides a unique application of the two concepts to topics spanning the social sciences and humanities"—Provided by publisher.
     ISBN 978–1–137–51474–5 (hardback)
     1. Biopolitics. 2. Utopias. 3. Biotechnology—Moral and ethical aspects. I. Stapleton, Patricia. II. Byers, Andrew.

JA80.B548 2015
321'.07—dc23   2014048055

A catalogue record of the book is available from the British Library.

Design by Newgen Knowledge Works (P) Ltd., Chennai, India.

First edition: June 2015

10 9 8 7 6 5 4 3 2 1

Contents

List of Figures and Tables vii

Permissions viii

Acknowledgments ix

Introduction 1
Andrew Byers and Patricia Stapleton

Part I Actions

1 American Bodies in a Time of War: The Militarized Body as a Utopian Space and Biopolitical Project for the State 13
Andrew Byers

2 "Abnormals" or "Exceptions": The Use of Technologies for Intersex People and People with Disabilities 41
Arpita Das

3 The Inauspicious Regulatory Beginnings of Preimplantation Genetic Diagnosis 63
Patricia Stapleton

Part II Speculations

4 Utopian Visions of "Making People": Science Fiction and Debates on Cloning, Ectogenesis, Genetic Engineering, and Genetic Discrimination 89
Evie Kendal

5 Decolonizing the Future: Biopolitics, Ethics, and Foresight through the Lens of Science Fiction 119
Selena Middleton

Part III Reactions

6 "All Day, All Week, Occupy Wall Street!": Space,
   Biopower, and Resistance   141
   Elena L. Cohen

7 Eating for the Future: Veganism and the Challenge of
   In Vitro Meat   167
   Rasmus R. Simonsen

Part IV Reflections

8 Utopia and Biopolitics: The Need for an Ethics in Biotechnology   193
   Cameron Barrows

List of Contributors   203

Index   207

Figures and Tables

Figures

6.1	Notice of temporary access change	145
6.2	Zuccotti Park	147
6.3	Homeland security intelligence document	148
7.1	"Disembodied Cuisine"—growing the steaks	173
7.2	"Disembodied Cuisine"	174
7.3	"Disembodied Cuisine"	174
7.4	"Disembodied Cuisine"—dinner	175
7.5	"Victimless Leather"	175

Tables

1.1	Overall draft and rejection statistics, by twentieth-century war	28
3.1	ART fertility clinic success rates reports for 2010, 2011, and 2012	66

Permissions

All images in chapter 7 "Eating for the Future: Veganism and the Challenge of In Vitro Meat" by Rasmus Simonsen, are courtesy of the Tissue Culture and Art Project (Oron Catts and Ionat Zurr). Hosted at SymbioticA, School of Anatomy and Human Biology, The University of Western Australia. The image credits are as under:

Title: 'Tissue Engineered Steak No.1' 2000 study for "Disembodied Cuisine"
Artists: The Tissue Culture and Art Project (Oron Catts and Ionat Zurr)
Medium: Pre-natal sheep skeletal muscle and degradable PGA polymer scaffold.
Date: 2000–01
Photography: Ionat Zurr
Explanatory information:
Part of Oron Catts and Ionat Zurr Research Fellowship in the Tissue Engineering and Organ Fabrication, MGH, Harvard Medical School.

Title: 'Tissue Engineered Steak No.1' 2000 study for "Disembodied Cuisine"
Artists: The Tissue Culture and Art Project (Oron Catts and Ionat Zurr)
Medium: Pre-natal sheep skeletal muscle and degradable PGA polymer scaffold, a micro gravity bioreactor.
Photography: Ionat Zurr
Date: 2000–01
Part of Oron Catts and Ionat Zurr Research Fellowship in the Tissue Engineering and Organ Fabrication, MGH, Harvard Medical School.

Title: Disembodied Cuisine Installation Nantes France 2003
Artists: The Tissue Culture and Art Project (Oron Catts and Ionat Zurr)
Date: 2003
Photography: Axel Heise

Acknowledgments

Andrew Byers: An excerpt of this chapter was presented at the 38th annual conference of the Society for Utopian Studies in Charleston, South Carolina, in November 2013. The author would like to thank the organizers and attendees of the conference as well as the following scholars who have kindly provided guidance and feedback on this project: Elizabeth Schreiber-Byers, Patricia Stapleton, Claire Curtis, Thomas Cassidy, Peter Stillman, Francesco Crocco, and Kenneth Roemer. Any mistakes in this chapter remain the sole responsibility of the author. Questions and comments are welcome; readers are invited to contact the author at andrew.byers@gmail.com.

Elena Cohen: The author would like to thank Rosalind Petchesky, Christina Quintana, Patricia Stapleton, and Andrew Byers for their guidance and feedback on this project. Any mistakes in this chapter remain the sole responsibility of the author. Questions and comments are welcome; readers are invited to contact the author at ecohen1@gradcenter.cuny.edu.

Evie Kendal: The author would like to acknowledge Catherine Mills, Andrew Milner, Olivia Khoo, and Ryan Tonkens for their valuable advice and input.

Patricia Stapleton: An excerpt of this chapter was presented at the 38th annual conference of the Society for Utopian Studies in Charleston, South Carolina, in November 2013. The author would like to thank the organizers and attendees of the conference, as well as the following people who have kindly provided guidance and feedback on this project: Andrew Byers, Claire Curtis, Kate Broad, Carrie Hintz, and Adam Luskin. Any mistakes in this chapter remain the sole responsibility of the author. Questions and comments are welcome; readers are invited to contact the author at pastapleton@wpi.edu.

Introduction

Andrew Byers and Patricia Stapleton

At its most literal, biopolitics is the merger of life with politics. Though the term biopolitics dates to the early decades of the twentieth century, it was only in the 1970s in the work of Michel Foucault that the idea of biopolitics, or "biopower" as he often described it, came to be understood as a fundamental and constitutive aspect of governance. Foucault defined biopower as the "explosion of numerous and diverse techniques for achieving the subjugations of bodies and the control of populations,"[1] as well as the set of mechanisms through which "the basic biological features of the human species became the object of a political strategy."[2] This refers not just to the power of life and death possessed by absolute monarchs in the premodern period, but also to the eventual control by even modern states and liberal democracies over how life within society would be lived and experienced.

Since the formation of the modern state system, governments have sought to exercise greater control over the lives and, increasingly, the bodies of their subjects and citizens. Indeed, entire disciplines of study—statistics, demography, epidemiology, and public health, among them—have arisen as highly politicized forms of knowledge, making possible the ability to better analyze and regulate ("govern") individuals and society as a whole and involving processes of "correction, exclusion, normalization, disciplining, therapeutics, and optimization."[3] Thus, the biological aspects of society, and the biology of the individual members of society, have come to be carefully managed by the state for its own ends. While biopolitical interventions by a state—and their most terrible effects on human bodies and lives—might be most visible in the large-scale eugenics programs practiced by authoritarian and liberal democracies alike in the nineteenth and twentieth centuries, contemporary areas of biopolitical concern are expansive, covering virtually the entirety of human existence, including reproduction, sexuality, physical and mental health, food consumption, appearance, and day-to-day activities, among many other areas. Indeed, biopolitics and biopolitical concerns play into all aspects of the functions of a state, from state formation to state maintenance and expansion; states have been and remain concerned with the management of populations to produce order and stability as well as managing an expanding population base from which state power can grow.

Over the course of the last two centuries, surveillance and disciplinary techniques—greatly aided by technological advances in the last several decades—have been developed for regulating the behaviors and bodies of ever larger populations with the goal of transforming them into "manageable subjects."[4] It is easy to perceive the disciplinary and surveillant nature of the modern state at work within the institutions of society where these are most explicit—prisons, schools, the military, for example—though as Gilles Deleuze pointed out, "societies of control" operate not simply by confining citizens in carceral institutions (though they do that too), but also through mechanisms of control permeated throughout society.[5] Control over human bodies, and therefore human behaviors and lives, is the means to a utopian end, in which these created "manageable subjects" will be cooperative, productive, and reproductive. To be clear, however, just as with other efforts at state control, individual and societal resistance to the imposition of large-scale biopolitical regulations is an important limiting factor on such efforts. Indeed, Foucault asserted the criticality of such resistance: "If there was no resistance, there would be no power relations. Because it would simply be a matter of obedience.... So resistance comes first, and resistance remains superior to the forces of the process; power relations are obliged to change with the resistance."[6] Thus, recognition of and resistance to biopolitical regulation can be profoundly transformative, producing new discourses, ideas, and social norms about biopolitics and the body.

Bodies are the mechanisms over which biopolitical power is exercised, and it is through the continual surveillance and disciplining of the body that the modern subject is subordinated to the state.[7] Before exploring these issues further, we should make clear that we do not take the term "body" as a self-evident concept. Sociologist Bryan Turner has described the body as "at once the most solid, the most elusive, illusory, concrete, metaphysical, ever present and ever distant thing—a site, an instrument, an environment, a singularity and a multiplicity."[8] We all have one, but grasping the nature of the body and making analytic and methodological sense of it is the challenge. It is important that we not simply invoke "the body" as a kind of generic object of study, as that tends to blur the distinctions between and among bodies; we need to ground bodies conceptually and historically, locating them spatially and temporally. As Elizabeth Grosz has pointed out, "bodies are never simply *human* bodies or *social* bodies [emphasis in original]."[9] With this in mind, Kathleen Canning has suggested that there are a number of different bodies to be investigated by scholars—among them, the social body as a whole; bodies as they are represented rhetorically and textually (and, we would add, visually); bodies as sites of experiences—for example, wounds and pregnancy—that indelibly shape them; and bodies as objects of regulation by the state and other institutions—all of which are explored in various ways by the authors in this collection.[10]

The biopolitical explorations in this volume fit within a larger literature; questions of the body and whether and how it may be regulated by the state have dominated modern politics for much of the twentieth century and into the present. In describing biopolitics, Foucault argued that "the disciplines of the body and the regulations of the population constitute the two poles around which the organization of power over life was deployed."[11] For Foucault, discipline is a mechanism of power that regulates the behavior—and the bodies—of individuals within the social or political body through regulation of space, time, and activities, enforced through a complex, almost ubiquitous system of surveillance, ultimately creating bodies that are at once docile, disciplined, and productive. This disciplinary power is part of what Foucault describes as "governmentality," in which officials and "experts" monitor, measure, and normalize individuals and entire populations through diffuse means, including public discourse, with an eye toward promoting particular models of health and personal behavior.[12]

In many ways, governmentality is the key linkage between biopolitics and utopia; it is "the ensemble formed by institutions, procedures, analyses and reflections, the calculations and tactics that allow the exercise of this very specific, albeit very complex power, which has as its target the population, as its major form of knowledge political economy, and as its essential technical means apparatuses of security."[13] Thus it is through governmentality that states strive to create the citizens best suited to fulfilling the utopian aspirations of the state, whatever those goals may be—for example, a productive, populous, physically fit citizenry compliant with the demands of the state. The state's acceptance of individuals as full citizens of society is highly contingent; factors as diverse as race, ethnicity, sexuality, and compliance with legalities, all play important roles in elevating or denying rights and freedoms to individuals. The consequences of biopolitical interventions and regulation are significant for individuals: not only do citizens become implicit or unwitting participants of the utopian experiment of the state, but they (and their bodies) also become regulated by the institutions and discourses of the state. Moreover, they are educated and trained throughout their lives to monitor and regulate their own behavior and bodies, in keeping with the goals of the state. Eventually monitoring and policing by state institutions could become almost secondary to self-monitoring and self-policing.

It is worth exploring exactly how and in what ways theorists have conceived of the idea of utopia as it applies to biopolitics in the modern world. Utopian scholarship encompasses a broad range of analytic approaches, and objects of study, which provides a great strength to the field, but a persistent problem in the field of utopian studies is the very definition of the term "utopia." Without such a consensus there can be little agreement on the very object of study.[14] Countless definitions have been offered. Ruth Levitas, one of the foremost utopian theorists, proposes a number of possibilities. She suggests that utopia might simply be defined as "how we would live and what kind of a world we would live in if we could do just that."[15] Levitas goes on to note that

> sometimes utopia embodies more than an image of what the good life would be and becomes a claim about what it could and should be: the wish that things might be otherwise becomes a conviction that it does not have to be like this. Utopia is then not just a dream to be enjoyed, but a vision to be pursued.... Those utopians who seek to make their dreams come true are deemed to be hopelessly unrealistic, or worse, actively dangerous.[16]

The chapters in this collection will examine a number of biopolitical utopian ideas that may have once seemed "hopelessly unrealistic," but with recent and near-future scientific developments, appear more realistic with every passing day. Many of these ideas, especially if put into practice, might well be considered "actively dangerous." In Levitas's most recent work on utopia, she goes on to describe utopia as perhaps being better framed as a method than as a goal, a method she describes as the "imaginary reconstitution of society."[17] For at least some, if not most, writers, philosophers, and theorists, such a conception of utopia likely does involve an imaginary reconstruction. The essays in this collection cover a broad range of utopian efforts and ideas, some of which do remain firmly in the realm of the imagination, while others, involving the efforts of some scientists, governments, and public policy advocates, range away from the strictly imaginary and into the realm of the real, or at least serious attempts to map utopian imaginings onto the real world, regardless of whether these ideas are ever fully realized.

In this volume, each author engages in some way with the utopian drive to control the body through the biopower of the state, to make those bodies reproductive, productive, disciplined, fit, homogeneous, normalized, or any other desirable set of traits, whether it is through a direct physical intervention or through the indirect training of citizens to instill self-regulatory practices. These biopolitical interventions may impact the individual body, the social body, or both, but all reveal the state's interest in perfecting its citizens. The chapters included in the volume cover a wide range of disciplines from across the social sciences and humanities: history, political science, women's and gender studies, bioethics, literature, and philosophy. Despite the diversity of topics represented by these fields, the authors reveal common characteristics in their treatment of utopia and biopolitics. Although links can be made among all of the chapters, we have organized the book into four sections to illustrate the main intersections of these two concepts.

In the first section, *Actions*, three chapters examine the practices of direct, medical intervention to "normalize" citizens' bodies. The individual body has become the site for such intervention because of some perceived failing, whether that be in physical ability, in reproductive capability, or in sexual characteristics. As a result, either through government effort or social pressure, individuals seek bodily "perfection." Andrew Byers addresses these issues in a historical review of the American government's efforts to engender a new American identity and conception of the body, one that would be healthy, physically fit, and thus capable of serving the nation militarily or otherwise in a time of war. He reviews the state's attempts to promote voluntary programs of physical fitness and their inherent utopian impulses. Byers concludes that the legacy of governmental interventions on American bodies is decidedly mixed, with the physical representations of bodies established during wartime having had a lasting impact on ideas about what constitutes an "ideal" body. Arpita Das also addresses societal stereotypes of what a "normal" body should be. To do so, she analyzes instances in which intersex and disabled bodies are categorized as "abnormal" and as "exceptions" in comparison to the normative body stereotype. Das shows how intersex and disabled bodies are alternately labeled as "weak" and as "threats," challenging the conception of what "normal" means. In contrast to the US government's efforts to prepare American bodies for military service, Patricia Stapleton's work reveals the absence of direct government regulation in the field of reproductive medicine. Yet, this absence demonstrates just as much a desire for perfecting potential American bodies as the clear efforts to increase the physical fitness of American citizens. Specifically, Stapleton's review of regulatory practices regarding preimplantation genetic diagnosis in fertility treatments illustrates how the lack of government regulation allows the "right" kinds of citizens to be born.

Speculations, the second section of the volume, pairs chapters that explore the intersection of utopia and biopolitics through a literary lens. Evie Kendal and Selena Middleton refer to science fiction texts in their respective works, and these references are employed as a starting point to examine how science fiction can reflect political and cultural fears about biotechnology. Both Kendal and Middleton are inherently dealing with the future tense, speculating over what new technologies might bring forth. Kendal explores the motivations for science fiction references in biopolitical debates, particularly in reference to genetic engineering in human biology. She finds that the present usage advances a socially conservative agenda because of the selected negative portrayals of the potential of new technologies; the most frequently used science fiction references are from technophobic cautionary tales that dramatize the threats posed to humanity. Similarly, Middleton looks at the fearful human response to future technologies presented in literature. She relies on the discourses of contemporary environmental movements and their concerns about genetically modified crops to analyze the fiction of Paolo Bacigalupi, finding that Bacigalupi's writing reflects the effects that social structures have on the role of science. Kendal and Middleton, respectively, call for the development of a positive relationship between the sciences and the humanities, suggesting that there is great potential in such a relationship to prepare society for the impacts of technological progress.

In the third section, *Reactions*, the authors turn to potential acts of resistance in the face of biopower. Elena Cohen explores the relationship between the Occupy Wall Street social movement and space, by looking at issues of resistance, power, security, sovereignty, discipline, governmentality, and biopolitics. She probes Foucault's concept of "heterotopia," building links between his conception of the term and the activities of the Occupy movement. Cohen concludes that the Occupy movement may be considered a hegemony-challenging heterotopia, providing resistance to the biopower of the state. In contrast to Cohen's focus on the citizen body and control over it through the regulation of space, Rasmus Simonsen turns our attention to the animal body and humans' relationship to meat. He focuses on a tenet of utopia that highlights freedom from scarcity in his treatment on veganism and the production of *in vitro* meat. Like the Occupiers challenging the status quo, Simonsen makes the case that veganism must continue to disrupt or deviate from the central social directionality of consuming animal products to remain effective.

Finally, the volume concludes with *Reflections*. Cameron Barrows offers an essay more philosophical in tone, requiring readers to reflect on the ethical dilemmas presented by modern-day science. He notes that the role of science since the Enlightenment has been focused upon the improvement and categorization of the body, and that science seems oblivious to its own utopian ideology. Barrows concludes that if biotechnology is part of our human future, we must develop an ethics determining the standards for its use.

Note that this collection is not comprehensive in terms of its coverage of biopolitical issues, nor is it intended to be. Rather, it attempts to sketch out some of the possibilities for how biopolitics and biopolitical debates have played out in the modern world, or could in the near future. For the most part, the essays in this collection deal with new and emergent areas of biopolitical interest (e.g., assisted reproductive technologies), many of which are enabled by recent and future scientific and technical developments, as well as areas that have not yet received sufficient scholarly attention (e.g., governmental promotion of fitness and health), rather than some of the most obvious forms of biopolitical intervention that have already received considerable analysis (e.g., eugenics). As such, this collection provides the reader with the opportunity to review some of the breadth of the actions, speculations, reactions, and reflections on contemporary biopolitical debates.

Notes

1. Michel Foucault, *The History of Sexuality, Volume 1: An Introduction* (New York: Vintage Books, 1990), 140. In addition to much of Foucault's œuvre, through which biopolitics is implicitly and explicitly present, Thomas Lemke's *Biopolitics: An Advanced Introduction* provides a useful overview and analysis of the concept of biopolitics. Thomas Lemke, *Biopolitics: An Advanced Introduction* (New York and London: New York University Press, 2011).
2. Michel Foucault, *Security, Territory, Population: Lectures at the Collège de France, 1975–76*, edited by François Ewald and Alessandro Fontana (Basingstoke and

New York: Palgrave Macmillan, 2007), 1. Foucault describes the birth of biopolitics in Michel Foucault, "17 March 1976 Lecture," in *"Society Must Be Defended!": Lectures at the Collège de France, 1975–76*, edited by Mauro Bertani and Alessandro Fontana (New York: Picador, 1997), 239–264.
3. Lemke, *Biopolitics*, 5.
4. Different segments of the population affected include: prisoners, hospital patients, students, workers, city dwellers, and so on. Jonathan Crary, *Techniques of the Observer: On Vision and Modernity in the Nineteenth Century* (Cambridge: MIT Press, 1990), 15.
5. Examples of the mechanisms of control include: networks of close-circuit television cameras in public spaces; electronic tags on people, animals, and goods; electronic health record management systems; continual assessment of performance in educational and professional settings; and continuous monitoring and training in workplaces. Gilles Deleuze, "Postscript on the Societies of Control," *October* vol. 59 (Winter, 1992): 3–7.
6. Michel Foucault, "Sex, Power, and the Politics of Identity," in *Ethics: Subjectivity and Truth: Essential Works of Michel Foucault*, Vol. 1 (New York: New Press, 1997), 167.
7. Michel Foucault, *Discipline and Punish: The Birth of the Prison*, 2nd Vintage ed. (New York: Vintage Books, 1995).
8. Bryan S. Turner, *The Body & Society: Explorations in Social Theory*, 3rd ed. (Los Angeles and London: SAGE, 2008), 43.
9. Elizabeth Grosz, *Space, Time, and Perversion: Essays on the Politics of Bodies* (New York: Routledge, 1995), 84.
10. Kathleen Canning, "The Body As Method? Reflections on the Place of the Body in Gender History," in *Gender History in Practice: Historical Perspectives on Bodies, Class & Citizenship* (Ithaca: Cornell University Press, 2006), 168–169.
11. Foucault, *The History of Sexuality*, 139.
12. For a concise summary of this concept, see Turner, *Body & Society*, 3–4.
13. Michel Foucault, "Governmentality," in *Understanding Foucault: A Critical Introduction*, 2nd ed., edited by Tony Schirato, Geoff Danaher, and Jen Webb (London: SAGE, 2012), 102.
14. Ruth Levitas, *The Concept of Utopia* (Syracuse: Syracuse University Press, 1990), 178.
15. Ibid., 1.
16. Ibid.
17. Ruth Levitas, *Utopia As Method: The Imaginary Reconstitution of Society* (Basingstoke and New York: Palgrave Macmillan, 2013), xi.

References

Canning, Kathleen. "The Body as Method? Reflections on the Place of the Body in Gender History." In *Gender History in Practice: Historical Perspectives on Bodies, Class & Citizenship*, 168–192. Ithaca: Cornell University Press, 2006.

Crary, Jonathan. *Techniques of the Observer: On Vision and Modernity in the Nineteenth Century*. Cambridge: MIT Press, 1990.

Deleuze, Gilles. "Postscript on the Societies of Control." *October* vol. 59 (Winter, 1992): 3–7.

Foucault, Michel. *The History of Sexuality, Volume 1: An Introduction*. New York: Vintage Books, 1990.

———. *Discipline and Punish: The Birth of the Prison*, 2nd Vintage ed. New York: Vintage Books, 1995.

———. "17 March 1976 Lecture." In *"Society Must Be Defended!": Lectures at the Collège de France, 1975–76*, edited by Mauro Bertani and Alessandro Fontana, 239–264. New York: Picador, 1997a.

———. "Sex, Power, and the Politics of Identity." In *Ethics: Subjectivity and Truth: Essential Works of Michel Foucault*, Vol. 1, 163–173. New York: New Press, 1997b.

———. *Security, Territory, Population: Lectures at the Collège de France, 1975–76*, edited by François Ewald and Alessandro Fontana. Basingstoke and New York: Palgrave Macmillan, 2007.

———. "Governmentality." In *Understanding Foucault: A Critical Introduction, Second Edition*, edited by Tony Schirato, Geoff Danaher, and Jen Webb. London: SAGE Publications Ltd, 2012.

Grosz, Elizabeth. *Space, Time, and Perversion: Essays on the Politics of Bodies*. New York: Routledge, 1995.

Lemke, Thomas. *Biopolitics: An Advanced Introduction*. New York and London: New York University Press, 2011.

Levitas, Ruth. *The Concept of Utopia*. Syracuse: Syracuse University Press, 1990.

———. *Utopia as Method: The Imaginary Reconstitution of Society*. Basingstoke and New York: Palgrave Macmillan, 2013.

Turner, Bryan S. *The Body & Society: Explorations in Social Theory*, 3rd ed. Los Angeles and London: SAGE Publications Ltd, 2008.

PART I

Actions

CHAPTER 1

American Bodies in a Time of War: The Militarized Body as a Utopian Space and Biopolitical Project for the State

Andrew Byers

Introduction: Biopolitics and American Politics

Contemporary American politics is, in many ways, biopolitics. The body itself, along with a host of imagined possibilities for it, lies at the center of many political struggles. It has become politicized, normalized, regulated, and controlled through technologies of surveillance and intervention. The American state has become increasingly interested in the regulation of human bodies in the areas of reproduction and health, highlighting the complex ways in which the state and society—the body politic—interact with the bodies of its citizens.[1] While we may think of this as a relatively recent development, the US government has conceived of the bodies of its citizens as a kind of biopolitical project, and as sites of intervention, throughout much of the twentieth century, first under the guidance of influential moral reformers during the First World War, and, beginning in the 1940s, as part of the creation of a national security state.[2] The 1940–41 discovery that half of all American men aged between 18 and 45 were medically unfit for military service ignited a debate about the health of the nation, involving discourses of the body among civilian and military leaders, physicians, educators, and fitness advocates lasting from 1940 to the 1960s. The Cold War brought renewed concerns when similar numbers of Korean and Vietnam War draftees were rejected for military service. The rapid rise of Soviet athletes in the Olympics and other international athletic competitions in the 1950s raised fears of a "muscle gap," which, according to at least some health and fitness advocates, might be as dangerous as the perceived missile gap.[3] The American body had always been imagined as strong, fit, and hardy, but this image was threatened when the true state of the health of American citizens was revealed, sparking concerns that Americans were growing soft, weak, flabby, and unfit. Even more alarmingly, the very health of the nation, and its continued ability to defend itself and battle its enemies abroad, seemed to many observers to also be endangered.[4]

Acknowledging the influence of Michel Foucault on our thinking about the body, Elizabeth Grosz notes that "power is inscribed on and by bodies through modes of social supervision and discipline as well as self-regulation."[5] We see this multipronged program to inscribe a regime of national fitness on American bodies through each of these means. Foucault's "docile bodies," pliable as objects of intervention and training, eventually function not just through external regulation and supervision, but are also increasingly made malleable and adapted via *self*-regulation and *self*-control; these eventually become not just docile bodies, but also docile wills, desires, and minds.[6] The inculcation of self-directed fitness goals is particularly relevant to the mid-century American case because of ongoing concerns that an overtly state-mandated fitness regime would meet resistance and be perceived as "un-American" due to its perceived intrusions into individual freedoms.

Foucault asserted that power does not simply "say no"—though at times it may indeed prohibit—but is instead productive, "engendering contextually-specific ways of knowing, being, feeling, acting, desiring... giv[ing] us our identities rather than denying them, distorting them, or taking them from us."[7] This kind of productive power works because it persuades individuals to take action and engage in new and modified behaviors, convincing them that doing so is in their best interest; self-surveillance becomes an important mechanism, particularly as a lighter mode of governance within a liberal democracy.[8] Governmental interventions via legal requirements and state mandates could run the risk of engendering resistance by citizens living in a society where they are not used to obvious, heavy-handed interventions by the state in intimate matters and identity formation. This is made evident in US government efforts that, with the cooperation of the medical profession and physical education community, sought to engender, via cooperative and voluntary means, an entirely new American identity and conception of the body, that of a healthy, physically fit body capable of serving the nation militarily or otherwise in a time of war, or during the Cold War, in a perpetual state of almost-war. This kind of biopolitical program marries several utopian impulses: a bodily utopia, a utopian conception of American citizenship, and a utopia of the newly reconstituted American nation. In the minds of the advocates of such an American utopia, the three were inextricably bound, with a rejuvenation of American bodies leading almost inevitably to transformation in American notions of citizenship and nation.

The bodies under discussion here are not simply the sites of intervention by an intrusive *welfare* state, as had been seen previously when social reformers and government officials sought to exercise their expanded reach and regulate the body to promote particular moral or hygienic visions in the early decades of the twentieth century. The bodies under discussion here are, in fact, sites of intervention by a growing *national security* state, beginning with World War II and continuing into the Cold War, that was less interested in endorsing a particular moral vision than in advocating a new, fit American body as demonstration of a strong, powerful, militarized American body politic that would be capable of defending the nation and aggressively asserting its interests against fascist or communist opponents.

While aspects of this debate centered on fears of what had become of American masculinity, concerns broadened to encompass fears about the physiques of the nation as a whole, to include not just men of military age, but women and children as well. The youth of America became an area of particular concern—and a site of tremendous interest for an imagined reconstruction—when it was demonstrated in 1953 that nearly 60 percent of American children could not meet minimum fitness standards.[9] Just as the US military had a massive pool of new recruits that it could modify and adapt to its own needs virtually at will, so too did schoolchildren provide a source of malleable bodies. National advertising campaigns, physicians, and the new group of fitness "experts" could urge the civilian parents of these children to get in shape, for their own good and the good of the nation, but such urgings could not readily be enforced in the home. Children, like military conscripts, represented a captive audience at school, however, where they could be made subject to direct intervention by teachers, coaches, doctors, and government officials. It is no coincidence that some of the most effective, and longest running, government programs designed to encourage physical fitness have targeted children.

World War II

But concerns about the fitness of American children came later; in the lead-up to American intervention in World War II, it was adult men of military age who initially represented the greatest immediate cause for concern. The nation's first peacetime draft, which began in 1940, offered public officials, physicians and health experts, physical fitness professionals, and other interested parties the opportunity to intervene in the lives—and bodies—of a growing pool of Americans as a means of resculpting the bodies of not just *soldiers* during wartime but also American *citizens* long after war's end. The American Association for Health, Physical Education, and Recreation's National Preparedness Committee asserted that the purpose of

> the nation is not merely to prepare to make war, it is to live before, during, and after. Many of the war-like virtues such as physical courage, stamina, cooperation, endurance, and faith in leaders are desirable traits in people at all times. Hence, plans for physical and social fitness must see beyond the immediate needs of combatant forces—important as these are—to the needs of youth in relation to national life.[10]

This vision of wartime opportunity to reshape the health of the nation in a postwar American society would capture the imagination of utopian-minded fitness advocates for the duration of the war.

As part of its effort to assess and induct large numbers of able-bodied men beginning in 1940 with the introduction of the draft, the army created an elaborate sorting and screening system for draftees. From October 1940 through the end of the war, Selective Service registered 49 million American men aged between 18 and 45 as potential draftees and examined approximately 18 million men between the ages of 18 and 37, the ages of men assessed as being most useful for the war effort.[11] Men were placed into four broad classes after their examination: Class I were deemed fit and available for military service; Class II were deferred from military service for up to six months because they had jobs deemed critical for the war effort; Class III were men with dependents; and Class IV were men who were *disqualified* for military service for a host of different reasons—for example, they might have already completed military service, or were conscientious objectors, or were considered "mentally, physically, or morally unfit" for military service after medical screening.[12] Hence, the infamous category of draftees classified IV-F, who failed to meet one or more of the physical or psychological requirements of military service. The terms "I-A" and "IV-F" passed almost immediately into common parlance, with one song, "Four-F Charlie" by Ted Courtney, describing such IV-F men as "complete physical wreck[s]" and "stout and always wheezing." The song went on to detail the failed masculinity—and impotence—of such men: "And his blood is thin as water, / He can never be a father."[13]

Of the 18 million men examined by the Selective Service System, more than 6.4 million (35.8 percent of the total) were rejected because they "were unable to meet the physical, emotional, mental, or moral requirements."[14] Two-thirds of those rejected were found to have one or more physical defects, 18.5 percent were said to have an emotional disorder, and 13.6 percent were said to have a mental or educational deficiency.[15] The War Department turned to its experiences in World War I to try to contextualize the high rejection rates of World War II.[16] In World War I, roughly four million men were examined by local selection boards and in the training camps to which those not immediately rejected were sent. Of these, more than 800,000 were rejected for all reasons (21.3 percent). Nearly all of them were rejected for physical defects. Just 25,000 men were rejected because they were said to be afflicted with an emotional disorder; another 43,000 were rejected for a mental or educational deficiency; and 19,000 men were rejected for "nonmedical" reasons, which in almost all cases meant that they were rejected on "moral grounds," for example, a known criminal record, an extremely bad personal reputation known to the local selection board, or if they were considered to be a "sexual deviant," among other reasons.

The comparison between the experiences of the two world wars made the World War II rejection rates seem even more anomalous. Not only were far more men, in an absolute and relative sense (21.3 percent vs. 35.8 percent), rejected for military service in World War II, but all rejection categories had increased, some dramatically. The rejection rate for physical defects had increased the least of all categories: 19.3 percent of men in World War II were rejected for physical reasons, though this still represented the largest category of rejections by far. The rate of rejection for emotional disorders, however, had increased by an order of magnitude—from 0.5 percent in World War I to 5.4 percent in World War II—and rejections for mental/educational deficiencies had increased from 0.9 percent to 4 percent (the "nonmedical"/moral rejection rate stayed virtually constant, increasing from just 0.4 percent to 0.5 percent). But these results are unsurprising given the changes the military made to its selection criteria in World War II: physical standards were toughened, as were educational requirements.[17] The psychological health of the nation had probably not declined markedly from one war to the next, but by World War II, the military had come to embrace the field of psychiatry and psychological screening standards, though it applied the latter unevenly to draft candidates, in ways that would have been inconceivable in World War I.[18] Moreover, this was, after all, a nation only beginning to recover from the abject poverty, deprivation, and poor nutrition brought about by the Great Depression.

The Roosevelt administration—and the nation as a whole—were alarmed to find that the bodies (and minds) of American men were not as healthy as they had been imagined to be. Army induction boards had expected to weed out just 2 percent of the draftees passed on to them by local draft boards, but instead were forced to reject 15 percent of draftees for failing the military's battery of screening tests. The War Department was even more alarmed when it discovered that local draft boards had already rejected 40 percent of prospective draftees for obvious physical reasons. Including the men rejected by local boards before they were even seen by the War Department, in all, more than half of American men called up to serve their country during World War II were rejected for military service.[19] The president stated that "he was worried about the health of the people of the United States as a result of the [rejection] figures."[20] Surgeon General Thomas Parran began calling the high rejection rate a "national disgrace."[21] Eleanor Roosevelt hoped that it would "give impetus to the movement for a comprehensive and nationwide health program."[22] The administration began to suggest that the high rejection rate was not an indicator of the poor health of the nation, but merely the product of extremely high standards set by the military. Colonel Leonard G. Rowntree, chief of the Selective Service System's Medical Division suggested that

> a rejection rate of 50 per cent means a curtailment in manpower of 50 per cent. It is not fair, however, to regard this evidence of invalidism or actual sickness in 50 per cent of the youth of the nation. Nor is it fair to accept it as indicative of 50 per cent of invalidism among the people in the country as a whole. It merely means that 50 per cent fail to attain the standard requirements of our Army.

Rowntree went on to add that "it is just and right also that we realize that the 50 per cent accepted, represents the finest manhood found in any army anywhere in the world today. Men must be really fit to serve in our army."[23]

During the war, it was not only the health and fitness of American men that was debated. Fitness advocates made clear that

> women, too, have a definite duty toward preparedness.... Women must take up much of the work of the world which was not formerly their

responsibility. This does not mean that the woman worker must become hard or must lose that femininity which makes her truly a woman. She must know that her physical and mental condition are part of the preparedness of the nation and that her health must be maintained in the best possible state.[24]

Reflecting the important role of women on the home front and increasing involvement of women in the workforce, particularly in critical defense industrial work, fitness advocates called for American women to improve strength, endurance, and teamwork, all of which, they suggested, could be addressed through greater involvement in athletic training programs.[25] One federal official noted that

> many people are still acting as though we could win the war with the armed forces and by male labor alone. The reader knows that we cannot. We have approximately 17 million women now employed. This is a larger figure than the combined military forces.... Women have put on the welder's mask, the steel helmet, safety shoes, and coveralls and have invaded the industrial plants. They are helping to make the planes, the ships, the tanks, the ammunition for the battlefront.... These women need a high level of fitness. They need strength, endurance, and body control.... Their fitness will help to bring our men back alive.[26]

Beyond improvements in fitness, medical interventions were urged for civilians as well, including a US Public Health Service (USPHS) campaign urging that all civilians, especially those working in war-related industries, have chest x-rays taken. One such USPHS poster read "You may look healthy, but what does your chest x-ray show?"[27]

Adult war workers were not the only targets for medical screening and health monitoring; schoolchildren also became subjects of medical inspections and classification. One such scheme proposed by Norman C. Wetzel, MD, in an article titled "The Simultaneous Screening and Assessment of School Children," suggested that children be subjected to a rigorous system of medical and fitness tests to measure their physiques and physical development over time. Using a classification system even more complex than that used by the military during the war, under Wetzel's plan, children would be sorted into various categories of health and fitness. The lowest category of children, the B5s, would correspond with the IV-Fs assessed by the military. This system would not only sort the bodies of children into categories of fitness, as with adult draftees, but would also serve as a kind of barometer for assessing the health of the nation.[28]

Fitness, as it was conceived during World War II, became an explicit duty of American citizens. The Selective Service advocated a national program of "prehabilitation...the correction of remediable defects before the prospective soldier reports to his local board for examination" and "rehabilitation...the correction of such disabilities after the registrant has been rejected by the local Selective Service Board or by the Induction Board of the army," in order to maximize the pool of available draftees who might be burdened with physical disabilities of a correctable nature.[29] While these efforts increased the number of initially rejected draftees who might later be rescreened and accepted for service, concerns about the health of the nation persisted for the duration of the war. By 1943, the Roosevelt administration was sufficiently concerned about the fitness of American civilians that it established the Committee on Physical Fitness, administratively housed under the Cabinet-level Federal Security Agency, to oversee such matters. The committee's responsibilities were to

> study problems relating to the promotion of physical fitness...and encourage the development of cooperative programs for their solution; serve as a center for the stimulation of state, district, and local programs for the promotion of physical fitness; make available to states, localities, organizations, and agencies upon request, the services of specialists in physical fitness; [and] prepare materials and serve as a clearing house on informational matters pertaining to the development of a national program of physical fitness.[30]

This wartime committee would become a model for similar organizations established by future presidents during the Cold War.

Many health and fitness advocates also used the opportunity of wartime to urge the creation and implementation of a national physical fitness program. "The objective," of this program, one proponent suggested,

> is to develop a strong, vigorous, and courageous people—a people with the bodily efficiency, skill, sentiment, and spirit to endure a long, hard war. The worth of this nation inevitably depends on that of its citizens and on their ability to endure the hardships of war and post-war reconstruction that confront us in the years ahead. Becoming "fit to fight and to serve on all fronts" is today the personal, patriotic duty of every American.[31]

The language of "total war,"[32] encompassing the home front as much as the fighting fronts, was common among advocates of national fitness, with even children conceived of as the next generation of soldiers and industrial workers:

> The total war in which we are engaged extends beyond the armed and industrial fronts. Of equal significance to these two fronts is the service on the educational and home fronts. Both in war and in peace, our youth of today become the men of tomorrow who must assume responsibilities for perpetuating our democratic ideals. To do so, they must be *physically and socially fit and must possess the will to serve* [emphasis in original].[33]

At the 1942 annual conference of the American Association for Health, Physical Education, and Recreation (AAHPER), association president Anne Schley Duggan argued that

> at home, as a sort of steady, sure basis for the total ... war in which we are engaged, we must have a strong-bodied, strong-spirited people characterized by strong morale. This means that we must also fight enemies other than the Axis powers. We must battle disease, weakness, debilitation, and all other factors which lessen the physical, mental, moral, and spiritual strength of our human resources—the most valuable and essential resources of all.[34]

The end of the war brought renewed reflection on American fitness efforts during the war and effects on the postwar nation. Fitness proponents congratulated themselves and their efforts for playing an important role in winning the war. As one such advocate put it,

> The sports program of American youth—baseball, football, basketball, soccer, track, and swimming—have given the American soldier an edge over the warriors of all nations.... The American soldier, both in World War I and World War II, demonstrated he had the necessary stamina and discipline to beat down the professional soldiers of Germany, Italy, and Japan; but in addition he possessed some things they lacked, namely, *initiative, originality, and the ability to think* [emphasis in original] in time of crisis. He had lived, worked, and played, both as an individual and as a member of a group, and in fighting, he followed true to form. The athletic programs of our schools and colleges have done as much to maintain a strong and sound America as any other single influence. These programs over the years have "democratized" individuals and groups to a degree little appreciated.... The sports program of America over the years has encouraged the making of Americans and the practice of democracy.[35]

Improved fitness, promoted through new public programs and sponsored by all levels of government, could thus provide the tools needed to win a war as well as improve American democracy as a whole.

During the course of World War II, it became clear that the new public and private emphasis on physical fitness was not just a matter of wartime expediency. There certainly was a strong desire to develop a larger pool of manpower for wartime military service, especially given the surprisingly large number of medical rejections of draftees in the first years of the draft, but the programs developed during the war—at least in the minds of its most vocal advocates—offered the opportunity to influence postwar conditions and transform the nation long after the war was over. In many ways, these writings reflect a kind of utopian desire to sustain and renew American patriotism by coupling personal and physical development with development on a national scale. Examples of this utopian impulse abound, coming across clearly in a variety of sources written by fitness advocates and health professionals in the public and private sectors. For example, Clyde E. Mullis advocated using a national youth swimming program to renew and reshape American patriotism and values of citizenship across a generation; Jane Cotton and Marjorie Wilson described a global civilization that narrowly escaped extinction in the war and the necessity of using health and physical education programs to create a "healthy, vigorous" postwar nation; and Clarence I. Chatto suggested that a renewed emphasis on public health and fitness would form the foundation for a postwar democratic society.[36] These and many other fitness and health advocates used the apparent national health crisis during wartime in an effort to begin reshaping the bodies, activities, and lifestyles of the nation's citizens with an eye toward crafting a new postwar vision of fitness for every American man, woman, and child.

The Cold War

Concerns about the nation's fitness did not abate at the war's end; as Cold War superpower competition intensified, so too did concerns about weak and flabby Americans, adults and schoolchildren alike. While wartime proposals for classifying children's fitness levels were never implemented, more than 4,200 American children were given the Kraus-Weber Minimal Fitness Tests, a set of exercises designed to test strength and flexibility, in the early 1950s. 57.9 percent of these children failed to meet even the minimum standard. Of the 2,900 European children of similar socioeconomic backgrounds who were administered the tests, only 8.7 percent did not meet the minimum requirements.

President Eisenhower was reportedly "shocked" by the results.[37] The media quickly took up the outcry over fears of growing "softness" of Americans, especially American schoolchildren. An article in the *U.S. News and World Report* asked "What's Wrong with American Youths?" going on to state "In terms of muscle and ability to do jobs requiring physical strength, the average American youth of today appears to be growing soft. His counterpart in some nations of Europe, enjoying fewer of the advantages of modern civilization, is stronger."[38] A later report on the Kraus-Weber Tests blamed American society's infatuation with leisure, luxury, and "push-button gadgets" for the disappointing results: "Our scientific and technological advances of today, while bringing an ease to living, deprive us of needed physical activity."[39] These Cold War–era fears of a growing "softness" of American youth reignited earlier debates about the health of the nation, reflected in a host of newspaper and magazine articles from the early 1950s with titles such as "Are We and Our Children Getting Too Soft?," "How Fit Are Our Children?," and "Why the President is Worried about Our Fitness."[40] That 52 percent of American men called up for military service during the 1950–57 draft were rejected only furthered these concerns.[41] During the Korean War, one of the most highly decorated Marines in US history, Lewis "Chesty" Puller, told his troops that "our country won't go on forever, if we stay soft as we are now. There won't be an America. Because some foreign soldiers will invade us and take our women and breed a hardier race."[42] Charles Brightbill, director of Defense Recreation Services at the Federal Security Agency, and former president of the American Recreation Society, argued at the height of the Korean War that civilian physical education programs should be reoriented to better prepare young male draftees for military service after high school:

> There needs to be a reexamination of school programs to the end that such programs are better equipped to prepare prospective military personnel for the recreational and physical aspects of service in the armed forces. Many military commanders complain that men coming into military service are virtually "recreational illiterates." They have neither the skills, the physical conditioning, nor the mental attitude [for military service].[43]

In response to concerns about the fitness of American youth in an era of Cold War confrontations, the Eisenhower administration established the President's Council on Youth Fitness (PCYF) in 1955.[44] The PCYF made clear that its promotion of fitness was inextricably bound up with the needs not of individuals but of the nation as a whole. As former heavyweight boxing champion Gene Tunney, who was involved with the PCYF in its early days, put it, "Anything we can do to direct the activities and energies of young America, in a wholesome and healthful way, is a measure of national defense."[45] Parents and children alike were continually reminded by officials and cooperative health and fitness experts that it was every American's civic duty to improve their fitness using a rhetoric that envisioned American children as future citizens, parents, and soldiers. The following year the administration sponsored the first President's Conference on Fitness of American Youth, held on June 18–19, 1956. Vice President Nixon gave the keynote address at the conference, warning that "we are not a nation of softies but we could become one, if proper attention is not given to the trend of our time, which is toward the invention of all sorts of gadgetry to make life easy and in so doing to reduce the opportunity for normal physical health-giving exercise."[46] The conference brought together 150 participants from the fields of sports, education, health, and related areas who came together to discuss youth fitness "in its broad meaning of total fitness, with implications for physical, mental, social, and spiritual well-being."[47]

Throughout its efforts, the PCYF continued to promote the concept of "total fitness," emphasizing not just *physical* fitness, but also "mental, moral, emotional, social, and cultural fitness."[48] As one total fitness advocate put it,

> many lay people and educators think of "fitness" in terms of physical fitness alone. This is especially true of physical educators, whose main emphasis is on that aspect of fitness, and on a vigorous physical conditioning program instead of a program of natural activity. We must proceed with caution along this course because physical fitness is not an end in itself but a means to an end—total fitness. This does not imply that physical fitness is undesirable—quite the contrary, it is a most desirable objective of our program, but it is not the only objective.[49]

The same total fitness proponent went on to say that "health, physical education, and recreation are not a separate area concerned with development of the physical body, but are a very important part of the entire educational program that is concerned with the whole development of the pupil—social, mental, emotional, spiritual, and physical."[50] The findings of the President's Conference on Fitness of American Youth, not surprisingly, echoed these concerns, noting that "physical fitness goes hand in hand with moral, mental, and emotional fitness" and that "a [comprehensive] fitness program is one which encompasses the total person—spiritual, mental, emotional, social, cultural, as well as the physical. Therefore, any stress on the physical element of youth development must be done in recognition of the interweaving of all personality factors."[51]

Despite calls for larger, more comprehensive—and compulsory—national fitness programs, the Eisenhower administration resisted, fearing that such a program smacked of totalitarianism, conjuring up images of Nazi Germany or the Soviet Union, and instead sought to create a distinctly *American* national fitness regime that would "sell the concept of fitness to the nation rather than force the issue."[52] To this end, it primarily relied on nongovernmental organizations like the US Olympic Committee, the Amateur Athletic Union, and the National Collegiate Athletic Association, along with individual school systems, to implement specific programs promoting athleticism and fitness.[53]

Throughout the 1950s, the concept of physical fitness was tightly linked with the kind of moral, psychological, and ideological fitness that Cold Warriors imagined was needed to stave off communist aggression. Senator George H. Bender (R-Ohio) explained to his constituents, "If we are going to retain our world leadership, we cannot permit this condition [lack of fitness] to continue. Mighty Rome collapsed when it lost its physical health. A flabby nation physically becomes a flabby nation mentally and spiritually."[54] In "Physical Fitness in the Pentomic Age," an article that could only have been written at the height of Cold War planning for how best to fight, and win, in a war with the Soviet Union, Simon A. McNeeley, a health specialist in the US Office of Education, noted that "in an age of atoms and missiles, mere man emerges as the ultimate weapon. The serviceman, now even more than in the past, is the backbone of a new 'pentomic' defense system, devised by our armed forces in the race to keep tactical doctrine in pace with scientific advances in the creation of deadly war machines."[55] The idea of the "Pentomic Age" referred to the new military organizational schema adopted by the US Army in 1957 in order to create units that would be more survivable and effective in a future war with the Soviets in which potentially large numbers of tactical nuclear weapons might be used on the battlefield. Noting that current and future American soldiers might be required to fight on a battlefield that had been saturated by atomic weapons, McNeeley pointed out the "greater physical demand upon the infantryman and his supporting troops" from such a kind of warfare, suggesting that the weapons of an atomic age of warfare would have "to be supplemented with the strength, agility, endurance, speed, and physical skills of the individual soldier."[56] For McNeeley, and presumably the US Office of Education (which would become the Cabinet-level US Department of Education in 1972), it was incumbent on school fitness programs to prepare young American men for such a possibility.

An article in the February 1952 issue of the *Journal of Health, Physical Education and Recreation* urged the development of healthy American citizens—the nation's most precious resource, it claimed—calling for the creation of a national fitness program for all students. That was not unusual, since many other fitness advocates had been suggesting such a program for the last decade. One of the subsidiary benefits of such a program, the article went on to explain, was that it might even include improvement in survival rates in the event of biological warfare and increased efficiency in civil defense programs.[57] Fit American bodies became one of the foundations of the increasingly emphasized American civil defense program of the 1950s.[58] It was not just the homes, schools, and workplaces of Americans that had to be strengthened and prepared to weather the effects of a Soviet nuclear strike, but the very bodies of the nation's civilians. "Good personal health is a basic essential for civil defense," stated the American Association for Health, Physical Education, and Recreation's (AAHPER) committee on education for civil defense.

> In addition to the health topics commonly taught under this heading, special emphasis should be given to: cleanliness; protection against the effects of atomic, biological, and chemical warfare; practical means of combating fatigue during emergencies; the structure and function of the human body as related to personal health; home care of the sick; importance of periodic health examinations; and the correction of remedial defects.[59]

The committee went on to state that

> there can be no real preparedness in the United States without adequate civil defense. This may be accomplished through a program of mutual protection on the part of individual persons, families, communities, and states. Effective civil defense requires increased educational effort to develop children and youth, as well as adults, into physically fit, mentally competent, emotionally stable, and morally sound American citizens.[60]

Civil defense advocates were careful to recommend that this kind of physical fitness training not take on too paramilitary a tone, however;

they consistently urged that Americans be strongly encouraged, but not compelled by the government to take part.

Concerns about the health and fitness of Americans in the Cold War only intensified under the Kennedy administration. Before he had even assumed office, President-Elect Kennedy published a widely read article in *Sports Illustrated* titled "The Soft American," in which he asserted that "we face in the Soviet Union a powerful and implacable adversary determined to show the world that only the Communist system possesses the vigor and determination necessary to satisfy awakening aspirations for progress." Kennedy, rhetorically at least, made American physical fitness crucial for waging the Cold War, going on to state that it might even "determine the future of freedom in the years to come.... Only if our citizens are physically fit will they be fully capable of such an effort."[61] Kennedy also raised the specter that not only was the fitness of the nation's young people poor, but that it was declining over time, citing the results of the annual physical fitness tests at Yale in which 51 percent of freshmen in 1951 had passed, but just 38 percent passed in 1960.[62] In fact, during the 1960 campaign, the decline in the nation's fitness and spirit became a weapon with which Kennedy bludgeoned his political opponents Nixon and Eisenhower. In a campaign speech titled "Are We up to the Task?" Kennedy asserted that Americans had "gone soft—physically, mentally, spiritually soft." Blaming the administration for the "erosion of our courage," he asserted that a younger, more vigorous president could restore and renew American fitness, and by extension, the nation's courage and the ability to withstand and triumph against the challenges of the Cold War.[63]

In the PCYF's first progress report undertaken by the victorious Kennedy administration, it recommended a plan of action to increase youth fitness that included identifying students with "a low level of muscular strength, agility, and flexibility" through a series of screening tests and that every public schoolchild be required to perform at least 15 minutes of "vigorous exercises and developmental activities" each day.[64] This plan of action was implemented, with some fanfare, under Kennedy. In 1963, the Kennedy administration changed the PCYF's name to the President's Council on Physical Fitness (PCPF) to enlarge the organization's scope, reflecting an interest in improving the physical fitness of adults and children alike, as well as a conceptual move away from play-oriented fitness programs to ones based on rigid, testable physical fitness standards.[65]

The Johnson administration continued to support the cause of physical fitness. In 1965, Johnson asserted that "the struggles to preserve

Table 1.1 Overall draft and rejection statistics, by twentieth-century war[a]

War	Draftees classified (thousands)	Draftees examined (thousands)	Draftees rejected (all reasons) (thousands)	% draftees examined who were rejected	Draftees inducted (thousands)
World War I	24,234	3,764	803	21.3	2,820
World War II	36,677	17,955	6,420	35.8	10,022
Korean War	9,123	3,685	1,189	32.3	1,560
Vietnam War	75,717	8,611	3,880	45.1	1,759

[a]Figures in table taken from Scott Sigmund Gartner, "Selected Characteristics of the Armed Forces–Personnel, Draftees, Medical Care, and Military Pay, by War: 1861–1975," Table Ed82–119 in *Historical Statistics of the United States, Earliest Times to the Present: Millennial Edition*, eds. Susan B. Carter, Scott Sigmund Gartner, Michael R. Haines, Alan L. Olmstead, Richard Sutch, and Gavin Wright (New York: Cambridge University Press, 2006).

N.B.: The rejection rates do not include those draft candidates who were rejected by local draft boards before they were turned over to the War Department, Selective Service System, or Defense Department. If those draftees were included, the total rejection rates would be even higher.

freedom and to advance human hopes and aspirations will not be won by nations whose citizens let themselves grow soft and weak.... Physical fitness is therefore a matter of national concern."[66] The next year, Johnson created the Presidential Physical Fitness Award, designed to spur physical fitness training and competition among the nation's schoolchildren. Given the massive escalation of American involvement in the Vietnam War under Johnson, it is no surprise that the health and fitness of American draftees was a major concern for his administration. Despite all the emphasis that had been placed on the nation's physical fitness from 1940 onward, the American draft-age male population showed little improvement. In 1964, 51 percent of all draft inductees examined for military service failed to meet physical or psychological standards.[67] By the second half of 1968, 45 percent of American men called up for preinduction military physicals were still being rejected for medical reasons, most commonly for obesity, a figure just slightly lower than the World War II rejection rate. A generation later, despite the new national emphasis on fitness and an attempt to create a new kind of American body for the Cold War, the athletic soldier-scientist-citizen, not much had changed. Almost half of American men were still deemed unfit to perform their imagined civic duty of military service (Table 1.1).[68]

Conclusion: Legacies of Biopolitical Interventions

What were the effects of this decades-long obsession with improving the health and fitness of the nation? Did the utopian imaginings of the possibilities for American bodies have any significant lasting impact, or were these simply products of wartime expediency and the peculiar era of the early Cold War? To be sure, there were some long-lasting effects, though the utopian hope of a renewed American society of fit citizens fell short of the mark. Governmental encouragement of physical fitness, with the cooperation of professional organizations like the American Medical Association, and schools across the nation, did help spur tremendous public interest in health, fitness, and nutrition. Without this public-private cooperation, the physical fitness movement might not have grown as quickly or to the same extent. The fitness and nutrition industries had been virtually nonexistent in 1940; they now bring in tens of billions of dollars each year, and the demand for these services only continues to grow, despite—or perhaps because of—the widespread prevalence of obesity and other chronic health concerns in America.

These activities set the stage for state intervention in the health and fitness of bodies, in ways large and small. The Obama administration's "Let's Move!" campaign—which some of its proponents have described in militarized fashion as part of a broader "war on childhood obesity"—would likely not have received popular support and acceptance had earlier administrations' programs not laid the groundwork.[69] Indeed, this most recent program is, in many ways, a repetition of earlier federal programs, none of which succeeded, in part, because of many Americans' resistance to pressure—whether that pressure be from governmental programs, society as a whole, or from corporate advertising campaigns—to transform their lifestyles and alter their levels of fitness and health via major diet and exercise changes. While the state may have sought to create and exploit the "docile" bodies of American citizens for its own ends, many of those individuals have strenuously resisted the calls for change. Short-term "fitness crazes" may have some moderate impact on the fitness of some members of society, but the lack of any long-term effects from these kinds of initiatives calls into question the ability of a state to effectively inculcate wholesale bodily transformation. By now most Americans know that they need to eat healthier and perform more exercise—the benefits for the individual and society alike are clear—but many are unwilling to make such changes in their personal lives, regardless of how persistently the government and other fitness advocates make the appeal.

The issues raised in the mid-twentieth century linking the physical fitness of adolescents with national security have hardly gone away. In the last few years, observers have frequently noted that roughly 75 percent of the nation's 17- to 24-year-olds—the military's prime targets for recruitments—are now ineligible for military service because of obesity, lack of educational attainment, serious criminal records, or physical ailments. Echoing the concerns of the 1950s, Curtis Gilroy, a Department of Defense official, stated that "when you get kids who can't do push-ups, pull-ups or run, this is a fundamental problem not just for the military but for the country." He went on to say that many young people are not "taking physical education in school; they're more interested in sedentary activities such as the computer or television. And we have a fast-food mentality in this country."[70] Mission: Readiness ("Military Leaders for Kids"), a group that describes itself as a "nonpartisan national security organization of senior retired military leaders calling for smart investments in America's children," has emerged to champion the cause of increasing childhood fitness for national security purposes.[71] The group's Executive Advisory Council is a virtual who's who of more than 300 retired generals and admirals. It lobbies policymakers and the public advocating changes in law and public policy to expand early childhood education programs, improve access to healthy food at public schools, and improve the quality and quantity of physical education for children. In April 2010, Mission: Readiness released its report "Too Fat to Fight," following it up two years later with "Still Too Fat to Fight." The initial report described childhood obesity as "an epidemic that threatens national security." The follow-up report went even further, stating: "The problem of junk food in schools is not just a national health issue. It is a national security issue."

The US government's attempts to address fears of declining health and fitness undoubtedly also represent a kind of militarization of American bodies, and the social body of the nation.[72] World War II had a tremendous impact on the American body, particularly the male draftees whose bodies were classified and categorized. These military screenings and the statistical data that emerged from them influenced later plans for rating the fitness of children, revamped physical education programs across the nation, led to industrial workers' health being monitored, and inspired efforts to strengthen the fitness of all Americans.[73]

Likewise, the programs and ideas described here established clear and long-lasting linkages between citizens' bodies, conscription and military service, and the very notion of citizenship. Conscription has long served not only as a solution to nations' needs for military manpower, but also as a mechanism for able-bodied, male citizens to publicly demonstrate their loyalty to the nation by providing military service. In December 1789 the French National Assembly stated this principle baldly: "Every citizen must be a soldier and every soldier a citizen or we shall never have a constitution."[74] When President Roosevelt signed into law the nation's first peacetime draft on September 16, 1940, he announced that by adopting a draft during peacetime, the United States had "broadened and enriched our basic concepts of citizenship. Besides the clear and equal opportunities, we have set forth the underlying other duties, obligations and responsibilities of equal service."[75] In both World War II and the Cold War, the bodies of American children—the bodies of future adult citizens coming of age in a time of national crisis—were perceived as sites of intervention and transformation. President Eisenhower later went on to urge that "our young people must be physically as well as mentally and spiritually prepared for American citizenship."[76] In a very real way, the bodies of the citizens called up to serve the nation in a time of crisis thus served as an extension of the national body politic, a reflection and barometer of the nation's health. Citizenship here is not merely a kind of legal or political categorization; it takes on a very tangible, physical component as well as an identity that must be socialized, trained, and practiced.[77] The state, along with partners in the medical and physical education communities, intervened in this socialization of citizenship through public education campaigns, physical training in schools, and medical inspections by physicians. It is important to note that while the socialization aspects and physical dimension of this notion of citizenship have the potential to create a unified conception of American citizenship and American bodies, they are also intensely *exclusionary*. What of the 50 percent of American men who cannot pass muster when called up for military service? What of those who suffer from uncorrectable medical problems and are unable to participate fully in physical education programs? Are these individuals still fully *citizens*, able to physically serve the nation, or are they somehow second-class citizens unable to perform the physical aspects of their civic duty?

The legacy of governmental interventions on the bodies of American citizens during wartime seems to have been decidedly mixed. This is a case where the truly personal became political. For those drafted and found physically unfit to serve their nation in time of war, the consequences were very real: they faced tremendous opprobrium and accusations of being both physically unfit and shirking their responsibilities as citizens and men, as well as economic penalties when they were refused jobs because of their draft status. For those whose bodies were *not* found wanting, many benefited from improved nutrition and exercise, while others had their bodies maimed or destroyed during war. Just as the state sought to benefit from the "improved" bodies of its citizens, so too did citizens have the opportunity to benefit from the attention and resources of the state allocated to improving public health. Physical representations of American bodies, particularly masculine ones, established during wartime, have had a lasting impact on our ideas about what constitutes an "ideal" body. Clearly able-bodiedness, along with youth and physical strength, all play important roles in these conceptions. This utopian model of what an American body should look like, and be capable of, has undoubtedly excluded millions of Americans who could not, literally, "measure up."

The case also raises questions about the desirability and extent of acceptable intervention by the state into matters of personal health. While some Americans seem content to have the state maintain a biopolitical interest in their personal health and fitness, largely accepting the notion that healthy, fit bodies are crucial to the nation during both war and peace, the repeated failure of such programs to effect large-scale changes to the bodies of American citizens suggests that a considerable amount of resistance exists. Indeed, the widespread resistance across the nation from school children, their parents, and public school officials to recent federally mandated changes to school lunch programs in the interest of reducing childhood obesity reinforces the idea that the state will continue to have great difficulty in convincing Americans to make even moderate changes to their lifestyles.[78] While the state has manifestly involved itself—along with the fitness and health industries—in creating a nearly omnipresent discourse surrounding physical fitness and health, with a goal of literally reshaping and reforming American bodies, it has not succeeded in actually improving the health of the nation. This suggests a serious limit to the utopian possibilities for a state when it encounters citizens who are not as docile as it might desire. Indeed, that every presidential administration from World War II to today has resisted imposing a fitness mandate for American citizens demonstrates an understanding that such a program would meet considerable resistance, and an acknowledgement of the problematic nature of utopian control of the body and the limitations of utopian aspirations in such fundamentally personal matters.

Notes

1. Bryan S. Turner, *Regulating Bodies: Essays in Medical Sociology* (London and New York: Routledge, 1992), 10, 46.
2. On World War I efforts to morally transform the nation via, in part, the regulation of bodies and sexuality, see Nancy K. Bristow, *Making Men Moral: Social Engineering during the Great War* (New York: New York University Press, 1996). On the creation of the American national security state, see Douglas Stuart,

Creating the National Security State: A History of the Law That Transformed America (Princeton: Princeton University Press, 2008) and Daniel Yergin, *Shattered Peace: The Origins of the Cold War and the National Security State* (Boston: Houghton Mifflin, 1977).
3. Jeffrey Montez de Oca, "The 'Muscle Gap': Physical Education and US Fears of Depleted Masculinity, 1954–1963" in *East Plays West: Sport and the Cold War*, edited by Stephen Wagg and David L. Andrews (New York: Routledge, 2007), 124–134; Shelly McKenzie, *Getting Physical: The Rise of Fitness Culture in America* (Lawrence: University Press of Kansas, 2013), 25. For examples of the literature in the popular press promoting the idea of a "muscle gap," see K. A. Cuordileone, *Manhood and American Political Culture in the Cold War* (New York: Routledge, 2005), 268n72.
4. McKenzie, *Getting Physical*, 17.
5. Elizabeth Grosz, "Inscriptions and Body-Maps: Representations and the Corporeal," in *Feminine/Masculine and Representation*, edited by Anne Cranny-Francis and Terry Threadgold (Sydney and Boston: Allen & Unwin, 1990), 65.
6. Grosz, *Space, Time, and Perversion*, 2.
7. Nikki Sullivan, "Foucault's Body," in *Routledge Handbook of Body Studies*, edited by Bryan S. Turner (Abingdon and New York: Routledge, 2012), 110.
8. Alexandra Howson, *The Body in Society: An Introduction*, 2nd ed. (Cambridge and Malden: Polity, 2013), 157–158.
9. Montez de Oca, "The 'Muscle Gap,'" 129–130.
10. "A Plan for National Preparedness through Health, Physical Education, and Recreation in Schools and Camps," *Journal of Health and Physical Education* (hereafter *JHPE*) (September 1940): 398.
11. Eli Ginzberg, *The Ineffective Soldier: Lessons for Management and the Nation, Volume 1: The Lost Divisions* (New York: Columbia University Press, 1959), 34–35; George Q. Flynn, *The Draft, 1940–1973* (Lawrence: University Press of Kansas, 1993), 60. After December 1942, men over the age of 37 were no longer accepted for induction.
12. Quoted in Christina S. Jarvis, *The Male Body at War: American Masculinity during World War II* (DeKalb: Northern Illinois University Press, 2004), 58.
13. Quoted in Jarvis, *The Male Body at War*, 60.
14. Ginzberg, *The Ineffective Soldier*, 35.
15. Ibid., 36.
16. There was significant public debate during the peacetime draft of 1940–41 (i.e., even before US entry into the war), about the significant number of rejections of draftees for physical defects. Comparisons with World War I were common, and arguments were made about why more men were being rejected in 1940–41 than in 1917–18 in order to reassure the public that physical standards were higher than previously, physical examinations more extensive, and the draftees being screened tended to be older than the prospective draftees of World War I. See for example, Amos R. Koontz, "Our Selectees Are Healthier!" *Hygeia* (December

1941): 954–956, 1007–1009. All data in the following paragraph taken from Ginzberg, *The Ineffective Soldier*, 142–143.
17. For example, in World War II men with only a few years of formal education, mostly working-class men from rural areas or recent immigrants, were much more likely to be rejected for educational reasons than in World War I, when 29 percent men of military age had six or fewer years of schooling. Ginzberg, *The Ineffective Soldier*, 137–166. Men sorted as I-A, were free from venereal disease, had excellent vision, possessed no missing limbs or digits, and were otherwise free from obvious medical problems, while men with relatively poor vision, missing digits, or infections with some venereal and other minor diseases, would have been accepted for service in World War I rather than be rejected outright or classified as fit for only limited service in World War II. Jarvis, *The Male Body at War*, 58–62.
18. Ginzberg, *The Ineffective Soldier*, 36–38, 142–144; Jarvis, *The Male Body at War*, 58–60, 75–77.
19. Flynn, *The Draft, 1940–1973*, 31–32; Jarvis, *The Male Body at War*, 60–61.
20. *New York Times*, October 3, 1941, quoted in Jarvis, *The Male Body at War*, 61.
21. Flynn, *The Draft, 1940–1973*, 32.
22. Quoted in Ibid., 32.
23. Leonard G. Rowntree, "National Rehabilitation," *JHPE* (April 1942): 223.
24. "Fit to Serve," *Hygeia* (August 1940): 679.
25. Paula D. Welch and Harold A. Lerch, *History of American Physical Education and Sport* (Springfield: C. C. Thomas, 1981), 179. See, for example, Elizabeth Halsey, "The Role of College Women in War," *JHPE* (May 1942): 283–284, 314 and Franz Schuck, "The Physical Education of Girls for the War Effort," *JHPE* (June 1943): 301, 345.
26. Dorothy La Salle, Senior Representative in Physical Fitness, Committee on Physical Fitness, Federal Security Agency "Fitness Today on the Home Front," *JHPE* (December 1944): 535.
27. Jarvis, *The Male Body at War*, 62–64.
28. Norman C. Wetzel, "The Simultaneous Screening and Assessment of School Children," *JHPE* (December 1942): 576–577, 622–624.
29. Leonard G. Rowntree, "Is Your Son Fit for Service?" *Hygeia* (January 1942): 39.
30. Executive Office, Committee on Physical Fitness, Federal Security Agency, "Civilian Physical Fitness," *JHPE* (December 1943): 518–519, 553–554. Executive Order 9338 established the Committee on Physical Fitness on April 29, 1943 under the leadership of John B. Kelly, chairman, and Paul V. McNutt, administrator. Its first meeting was held on June 16, 1943, and it operated for the duration of the war.
31. Hiram A. Jones, "Report on National Fitness: A Program through Schools and Colleges," *JHPE* (March 1942): 134.
32. There is a rich literature on the idea of total war, the total mobilization of society for war, as well as the use of widespread violence across the civil-military divide, with World War II inevitably cited as the purest example of such a war. See for example, Roger Chickering, Stig Förster, and Bernd Greiner (eds.), *A World At*

Total War: Global Conflict and the Politics of Destruction, 1937–1945 (Washington and Cambridge: German Historical Institute and Cambridge University Press, 2005).
33. Jones, "Report on National Fitness," 135.
34. Anne Schley Duggan, "Presentation of the Conference Theme," *JHPE* (June 1942): 332. The theme of the 1942 AAHPER national convention was "National Fitness through Health, Physical Education, and Recreation—Fitness for Victory."
35. John S. Herron, "Human Relations, Democracy and Physical Education," *JHPE* (November 1946): 511.
36. Clyde E. Mullis, "Developing Citizenship through a Swimming Program," *JHPE* (June 1944): 310–311, 342; Jane Cotton and Marjorie Wilson, "Postwar Planning in Physical Education," *JHPE* (October 1944): 435–436, 468; and Clarence I. Chatto, "Health and Physical Education for Democratic Living," *JHPE* (October 1946): 466–467, 498–499.
37. Matthew T. Bowers and Thomas M. Hunt, "The President's Council on Physical Fitness and the Systematisation of Children's Play in America," *The International Journal of the History of Sport* 28:11 (2011): 1496–1511; McKenzie, *Getting Physical*, 14–15; Montez de Oca, "The 'Muscle Gap,'" 129.
38. *US News and World Report* (March 19, 1954): 35. The media outcry over the test results is discussed in Marjorie Phillips, "How Fit Are Our American Schoolchildren?," *JHPE* (September 1955): 14–15, 71. In the same article Phillips noted that some investigators at Indiana University questioned the methodology of the Kraus-Weber Tests, which must have offered some assurances to those concerned with the growing softness of American youths.
39. Quoted in McKenzie, *Getting Physical*, 41. This theme of increased use of new technologies causing poor physical fitness was a common one throughout the era, and was picked up again by President-Elect Kennedy in 1960: "Many of the routine physical activities which earlier Americans took for granted are no longer part of our daily life. A single look at the packed parking lot of the average high school will tell us what has happened to the traditional hike to school that helped to build young bodies. The television set, the movies and the myriad conveniences and distractions of modern life all lure our young people away from the strenuous physical activity that is the basis of fitness in youth and in later life." John F. Kennedy, "The Soft American," *Sports Illustrated* (December 26, 1960): 15–17.
40. McKenzie, *Getting Physical*, 15. See also representative pieces like Laurence E. Morehouse, "American Living—A Threat to Fitness," *JHPE* (September 1956): 20, 69.
41. McKenzie, *Getting Physical*, 43–44.
42. Quoted in Montez de Oca, "The 'Muscle Gap,'" 125.
43. Charles Brightbill, "Youth in the Armed Forces," *JHPE* (April 1951): 10–11.
44. The President's Council on Youth Fitness was established on July 16, 1956 by Executive Order 10673. McKenzie, *Getting Physical*, 14–53; Montez de Oca, "The 'Muscle Gap,'" 129–134; Thomas M. Hunt, "American Sport Policy and

the Cultural Cold War: The Lyndon B. Johnson Presidential Years," *Journal of Sport History* 33 (2006): 273–297.
45. Quoted in McKenzie, *Getting Physical*, 25. This anxiety about growing physical weakness due to social and cultural changes brought about by the availability of new "luxury" technologies was also present during World War II, though it took on renewed urgency during the 1950s. Tunney had been one of the most vocal advocates for increased physical fitness for American children even during World War II. In 1942, Tunney worried that "We no longer think of going out for long hikes on days off; the most popular pastime is sitting in the movies, or listening to the radio. Our forms of entertainment have changed and because of that we have suffered a physical reaction. The youngster of today doesn't have the opportunity to walk to school, or to chop wood, or stoke furnaces, or even climb stairs. He now has elevators or escalators; furnaces are automatically operated, and he has a bus to carry him to the very steps of school." Tunney quoted in William M. Harlow, "Organized Camping and Defense," *JHPE* (April 1942): 231.
46. "The President's Conference on Fitness of American Youth," *JHPE* (September 1956): 8.
47. Ibid., 9.
48. Quoted in McKenzie, *Getting Physical*, 11.
49. Gene Kidder, "All-Around Fitness for All," *JHPE* (September 1957): 8.
50. Ibid.
51. "Fitness of American Youth: Report on the President's Conference on Fitness of American Youth," *JHPE* (March 1957): 33.
52. McKenzie, *Getting Physical*, 24, 26.
53. Montez de Oca, "The 'Muscle Gap,'" 124.
54. Quoted in McKenzie, *Getting Physical*, 41.
55. Simon A. McNeely, "Physical Fitness in the Pentomic Age," *JHPE* (September 1958): 21.
56. Ibid.
57. "Mobilization News: Planning for Civil Defense in Health, Physical Education, and Recreation Programs," *JHPE* (February 1952): 47.
58. For a good discussion of the US civil defense program during the Cold War, see Laura McEnaney, *Civil Defense Begins at Home: Militarization Meets Everyday Life in the Fifties* (Princeton: Princeton University Press, 2000).
59. "Mobilization News: Education for Civil Defense," *JHPE* (February 1953): 37.
60. Ibid., 38.
61. John F. Kennedy, "The Soft American," *Sports Illustrated* (December 26, 1960): 15–17. As president, Kennedy went on to publish a second article, "The Vigor We Need," *Sports Illustrated* (July 16, 1962): 12–15. There is a certain amount of irony in the fact that John F. Kennedy, like his predecessor Franklin D. Roosevelt, had been plagued by ill health for much of his life, though Kennedy carefully cultivated a healthy, vigorous, athletic public persona, despite the debilitating and chronic pain from which he suffered, and kept from public knowledge.
62. John F. Kennedy, "The Soft American," 15–17.

63. Quoted in Cuordileone, *Manhood and American Political Culture in the Cold War*, 173.
64. "A Progress Report on the President's Fitness Program," *JHPE* (September 1961): 31.
65. Bowers and Hunt, "The President's Council on Physical Fitness and the Systematisation of Children's Play in America," 1501.
66. Quoted in Hunt, "American Sport Policy and the Cultural Cold War: The Lyndon B. Johnson Presidential Years," 280.
67. Flynn, *The Draft, 1940–1973*, 207.
68. Ibid., 180.
69. The Obama administration's Let's Move! campaign has not been without controversy; critics of the administration, the campaign itself, and Michelle Obama have variously suggested that the federal government should not intervene in the dietary and fitness activities of the nation's schoolchildren, and it has been widely reported that the meals offered under the program have been unpopular with children.
70. Christian Davenport and Emma Brown, "Girding for an Uphill Battle for Recruits: Obesity, Poor Education Big Obstacles to Military Recruiting," *Washington Post* (November 5, 2009).
71. For more on Mission: Readiness, as well as copies of all its official reports, see http://www.missionreadiness.org/ (accessed May 27, 2014).
72. Montez de Oca, "The 'Muscle Gap,'" 134–135.
73. Jarvis, *The Male Body at War*, 5.
74. Quoted in Flynn, *The Draft, 1940–1973*, 2.
75. Ibid.
76. Quoted in Montez de Oca, "The 'Muscle Gap,'" 134.
77. Ibid., 123.
78. Kay E. Brown, "School Lunch Modifications Needed to Some of the New Nutrition Standards," Testimony Before the Subcommittee on Early Childhood, Elementary, and Secondary Education, Committee on Education and the Workforce, House of Representatives, US Government Accounting Office (GAO) Report GAO-13-708T (June 27, 2013).

Bibliography

Bowers, Matthew T., and Thomas M. Hunt. "The President's Council on Physical Fitness and the Systematisation of Children's Play in America." *The International Journal of the History of Sport* 28.11 (2011): 1496–1511.

Brightbill, Charles. "Youth in the Armed Forces." *Journal of Health and Physical Education* (April 1951): 10–11.

Bristow, Nancy K. *Making Men Moral: Social Engineering during the Great War.* New York: New York University Press, 1996.

Brown, Kay E. "School Lunch Modifications Needed to Some of the New Nutrition Standards," Testimony Before the Subcommittee on Early Childhood,

Elementary, and Secondary Education, Committee on Education and the Workforce, House of Representatives, US Government Accounting Office (GAO) Report GAO-13–708T (June 27, 2013).

Chatto, Clarence I. "Health and Physical Education for Democratic Living." *Journal of Health and Physical Education* (October 1946): 466–467, 498–499.

Chickering, Roger, Stig Förster, and Bernd Greiner (eds.). *A World at Total War: Global Conflict and the Politics of Destruction, 1937–1945*. Washington and Cambridge: German Historical Institute and Cambridge University Press, 2005.

Cotton, Jane, and Marjorie Wilson. "Postwar Planning in Physical Education." *Journal of Health and Physical Education* (October 1944): 435–436, 468.

Cuordileone, K. A. *Manhood and American Political Culture in the Cold War*. New York: Routledge, 2005.

Davenport, Christian, and Emma Brown. "Girding for an Uphill Battle for Recruits: Obesity, Poor Education Big Obstacles to Military Recruiting." *Washington Post* (November 5, 2009).

Duggan, Anne Schley. "Presentation of the Conference Theme." *Journal of Health and Physical Education* (June 1942): 332.

Executive Office, Committee on Physical Fitness, Federal Security Agency. "Civilian Physical Fitness." *Journal of Health and Physical Education* (December 1943): 518–519, 553–554.

"Fitness of American Youth: Report on the President's Conference on Fitness of American Youth." *Journal of Health and Physical Education* (March 1957): 33.

"Fit to Serve." *Hygeia* (August 1940): 679.

Flynn, George Q. *The Draft, 1940–1973*. Lawrence: University Press of Kansas, 1993.

Gartner, Scott Sigmund. "Selected Characteristics of the Armed Forces—Personnel, Draftees, Medical Care, and Military Pay, by War: 1861–1975," Table Ed82–119. In *Historical Statistics of the United States, Earliest Times to the Present: Millennial Edition*, edited by Susan B. Carter, Scott Sigmund Gartner, Michael R. Haines, Alan L. Olmstead, Richard Sutch, and Gavin Wright (New York: Cambridge University Press, 2006).

Ginzberg, Eli. *The Ineffective Soldier: Lessons for Management and the Nation, Volume 1: The Lost Divisions*. New York: Columbia University Press, 1959.

Grosz, Elizabeth. "Inscriptions and Body-Maps: Representations and the Corporeal." In *Feminine/Masculine and Representation*, edited by Anne Cranny-Francis and Terry Threadgold. Sydney and Boston: Allen & Unwin, 1990.

Halsey, Elizabeth. "The Role of College Women in War." *Journal of Health and Physical Education* (May 1942): 283–284, 314.

Harlow, William M. "Organized Camping and Defense." *Journal of Health and Physical Education* (April 1942): 231.

Herron, John S. "Human Relations, Democracy and Physical Education." *Journal of Health and Physical Education* (November 1946): 511.

Howson, Alexandra. *The Body in Society: An Introduction*, 2nd ed. Cambridge and Malden: Polity, 2013.

Hunt, Thomas M. "American Sport Policy and the Cultural Cold War: The Lyndon B. Johnson Presidential Years." *Journal of Sport History* 33 (2006): 273–297.
Jarvis, Christina S. *The Male Body at War: American Masculinity during World War II*. DeKalb: Northern Illinois University Press, 2004.
Jones, Hiram A. "Report on National Fitness: A Program through Schools and Colleges." *Journal of Health and Physical Education* (March 1942): 134–135.
Kennedy, John F. "The Soft American." *Sports Illustrated* (December 26, 1960): 15–17.
———. "The Vigor We Need." *Sports Illustrated* (July 16, 1962): 12–15.
Kidder, Gene. "All-Around Fitness for All." *Journal of Health and Physical Education* (September 1957): 8.
Koontz, Amos R. "Our Selectees Are Healthier!" *Hygeia* (December 1941): 954–956, 1007–1009.
La Salle, Dorothy. "Fitness Today on the Home Front." *Journal of Health and Physical Education* (December 1944): 535.
McEnaney, Laura. *Civil Defense Begins at Home: Militarization Meets Everyday Life in the Fifties*. Princeton: Princeton University Press, 2000.
McKenzie, Shelly. *Getting Physical: The Rise of Fitness Culture in America*. Lawrence: University Press of Kansas, 2013.
McNeely, Simon A. "Physical Fitness in the Pentomic Age." *Journal of Health and Physical Education* (September 1958): 21.
"Mobilization News: Education for Civil Defense." *Journal of Health and Physical Education* (February 1953): 37.
"Mobilization News: Planning for Civil Defense in Health, Physical Education, and Recreation Programs." *Journal of Health and Physical Education* (February 1952): 47.
Montez de Oca, Jeffrey. "The 'Muscle Gap': Physical Education and U.S. Fears of Depleted Masculinity, 1954–1963." In *East Plays West: Sport and the Cold War*, edited by Stephen Wagg and David L. Andrews. New York: Routledge, 2007.
Morehouse, Laurence E. "American Living—A Threat to Fitness." *Journal of Health and Physical Education* (September 1956): 20, 69.
Mullis, Clyde E. "Developing Citizenship through a Swimming Program." *Journal of Health and Physical Education* (June 1944): 310–311, 342.
Phillips, Marjorie. "How Fit Are Our American Schoolchildren?" *Journal of Health and Physical Education* (September 1955): 14–15, 71.
"A Plan for National Preparedness through Health, Physical Education, and Recreation in Schools and Camps." *Journal of Health and Physical Education* (September 1940): 398.
"A Progress Report on the President's Fitness Program." *Journal of Health and Physical Education* (September 1961): 31.
Rowntree, Leonard G. "Is Your Son Fit for Service?" *Hygeia* (January 1942): 39.
———. "National Rehabilitation." *Journal of Health and Physical Education* (April 1942): 223.
Schuck, Franz. "The Physical Education of Girls for the War Effort." *Journal of Health and Physical Education* (June 1943): 301, 345.

Stuart, Douglas. *Creating the National Security State: A History of the Law That Transformed America.* Princeton: Princeton University Press, 2008.

Sullivan, Nikki. "Foucault's Body." In *Routledge Handbook of Body Studies*, edited by Bryan S. Turner. Abingdon and New York: Routledge, 2012.

"The President's Conference on Fitness of American Youth." *Journal of Health and Physical Education* (September 1956): 8.

Turner, Bryan S. *Regulating Bodies: Essays in Medical Sociology.* London and New York: Routledge, 1992.

Welch, Paula D., and Harold A. Lerch. *History of American Physical Education and Sport.* Springfield: C. C. Thomas, 1981.

Wetzel, Norman C. "The Simultaneous Screening and Assessment of School Children." *Journal of Health and Physical Education* (December 1942): 576–577, 622–624.

"What's Wrong with American Youths?" *U.S. News and World Report* (March 19, 1954): 35–36.

Yergin, Daniel. *Shattered Peace: The Origins of the Cold War and the National Security State.* Boston: Houghton Mifflin, 1977.

CHAPTER 2

"Abnormals" or "Exceptions": The Use of Technologies for Intersex People and People with Disabilities

Arpita Das

Bodies that are intersex or physically disabled do not conform to socially constructed normative standards. Ideas of normalization pressure people into attempting to achieve a certain kind of body through a variety of measures such as corrective surgeries, rigorous fitness regimes, and medication, among others. Contemporary medicine harbors the notion of the body as an object to be modified and controlled, thus evoking the idea of a completely utopian body that is indestructible, regenerative, and capable of self-healing. This utopian projection of the perfect body is based on the idea of a body with deficits or defects, but whose functions can be realized at an optimal level because of a belief in the possibilities of improvement.[1] In other words, even if the body has defects, they could be corrected and improved upon with medical treatments and surgery in a utopian world. Both feminists and disability theorists have critiqued the "normative" body that is idolized by popular media as well as by many health specialists.[2] Garland-Thomson advocates for a feminist disability theory that can help in critiquing interventions that normalize the nonstandard body.[3] For example, in many cases, intersex children are "normalized" through surgery and "correction" right after birth. People with physical disabilities, too, often undergo a number of interventions through their lifetime in order to fit the normative body stereotypes.

In this chapter, I focus on how certain bodies such as those that are intersex or physically disabled are considered either as "abnormal" or as the "exception," and how various technologies are used to make these bodies conform to the idea of the "normal." In certain situations, however, technology can contribute to the further distancing of these bodies from the "normal," even rendering these bodies as threats. In other situations, technology can create bodies that do not conform to normative standards, but yet they may be found desirable by the state, such as in competitive sports. Thus, I explore how disabled and intersex bodies are rendered *abnormal* and are therefore constructed as in need of "correction" within the biopolitical framework as discussed by Foucault[4] and further expanded on by Rose.[5] I use a biopolitical framework as my analytic lens, in addition to examining some of the intersections between intersexuality and disability, to illustrate the varying definitions, narratives, and effects of "normality" on both society and the state.

Theoretical Underpinnings

Within the biopolitical framework, as discussed by Foucault,[6] the state extends power over the physical and political bodies of its population. During the seventeenth and particularly the eighteenth centuries, the state shifted its attention from individual bodies to the population as a whole and therefore the government found it important to measure matters of fertility, mortality, morbidity, and health status of the population. The state was concerned not so much from the perspective of the health or the rights of the population, but was instead interested in how the population performed in matters of production and reproduction. A robust and good health status of the population would lead to better production. Also, a "healthy" population would help in reproducing healthy babies who would, in turn, form a good workforce in the future. Even after these measures were put in place, the state did not stop paying attention to individual bodies; it focused on the general population in addition to its regulation of individual bodies.

The government's attention to individual bodies focused on rendering certain bodies as "normal," leaving others to be subjected to normalization procedures.[7] Bodies are rendered "abnormal" when they do not adhere to societal notions of an "able" and "functioning" body. These bodies could differ in terms of their appearance and ability to (re)produce, among other things. The procedures of normalization are not restricted to the level of limbs and organs (molar level), but they also address more minute levels of genes and chromosomes (molecular level).[8] Consequently, bodies could be "corrected" and "altered" at both the molar and molecular levels. "Abnormal" bodies are considered non-normative and not adding to the nation's worth. Nonetheless, there are also *exceptions*, where both intersex people and people with disabilities are considered not as weak and dependent but instead as a threat to the nondisabled body.

As Agamben[9] has discussed, exceptions redefine the law and itself; whereas law can operate only in general situations (without chaos), it is only in the case of exceptions or extraordinary circumstances, that the general situation, the law, and the exception itself get defined. In this chapter, I consider competitive athletic sports as an arena where the idea of the "exceptions" among intersex bodies and physically disabled bodies play out. First, in the case of intersex people, sometimes the so-called abnormality or the intersexuality of the person is seen as a threat in competitive games when they are observed to perform better than socially constructed normative females. In these cases, the intersex people are then prescribed medical and other procedures to mitigate and reduce their "abnormality." Alternatively, people with disabilities may pose a threat when they are perceived to out-perform the nondisabled population in competitive games when using prosthetic technologies. For example, the use of artificial limbs in competitive athletics by Oscar Pistorius, a famous South African runner, not only allowed him to compete with the mainstream nondisabled athletes, but also cast him as a threat. In this situation, his disability and the use of technology to compensate for his disability, which was considered to be an unfair advantage, became a threat for the rest.

These cases make for an interesting study of how abnormality is construed in exceptional circumstances: redefining the norm while at the same time deeming these bodies as "abnormal." These instances also illustrate how so-called abnormal bodies that are disabled or do not fit into the male-female binary present themselves as sites of resistance against stereotypical societal notions of being weak, unruly, and unproductive, thereby posing important questions about stereotypes. It is often this resistance that provokes anxieties among societies and systems that manifest in binary categories of male-female, able-disabled, and normal-abnormal while simultaneously disassembling these simplistic classifications.

Guided by notions of correction and alteration to these "abnormal" bodies, both disabled bodies and intersex bodies are frequently subjected by society to medical management. These alterations may also be made at the level of genes through processes of genetic engineering, sometimes bordering on eugenics. These corrections are aimed not just at children post-birth, but also on fetuses yet to be born, by making alterations in the prenatal environment or the genetic makeup of the fetus.

While there exist multiple and conflicting definitions of intersexuality, in this chapter I use the term "intersex people" to signify people who "are born with physical, hormonal or genetic features that are neither wholly female nor wholly male; or a combination of female and male; or neither female nor male."[10] Therefore, intersex people do not strictly adhere to being either male or female. This nonadherence may happen at different levels: at the level of the external genitalia (in which case, the intersexuality is detected earlier), at the level of internal reproductive organs, and at the level of hormones and chromosomes. Many people may go through their entire lives without being aware of their intersexuality. Intersexuality is localized in the body and does not have a bearing on one's gender identity. Intersex people are often conflated with transgender people and, although there exist some similarities at the level of discrimination and stigma, they are disparate entities. For the purposes of this chapter, I restrict my analysis to intersex people alone.

Like intersexuality, disability is not a homogenous or monolithic category; it consists of many variations. People may have various kinds of disabilities: physical and intellectual. Some people may be born with a disability and others may acquire it during their lifetime. Some disabilities may be temporary and can be treated, while others may not be. Although there is some activism, advocacy, and interventions for people with physical disabilities, people with intellectual disabilities remain largely ignored. In this chapter, I focus on people with physical disabilities alone, as I explore the intersections between people with disabilities and intersex people at the level of the body.

As a caveat, it is important to mention that in exploring the intersections between intersex people and people with disabilities, my aim is not to conflate the two. In discussing the two together, I do not label intersex people as "disabled" nor do I advocate doing so. I am aware of the debates within the intersex movement that wish to steer clear of being categorized as "disabled" not only because of the additional stigma of disability, but also because the two categories are different and must be treated so. In this chapter, I merely attempt to look at how both groups are often treated as the "abnormals" within the biopolitical framework, in need of and subjected to correction, and how the use of technologies can impact these two different communities in similar ways to make them the exceptions.

Discourses of "Abnormality"

Both intersex bodies as well as disabled bodies are considered atypical and "abnormal" as per societal norms. These norms accord more privilege to the so-called normal bodies while dis-privileging and marginalizing others. According to Garland-Thomson,[11] the ability-disability system works through differentiating and marking bodies, resulting in an unequal distribution of resources and power within the society.

The normalization of bodies happens not only at the levels of ability or sex; raced and different bodies are also rendered "abnormal" and subjected to public scrutiny. Historically, the public exhibition of Saartje Baartman, also referred to as the "Hottentot Venus," exemplifies this phenomenon. Saartje, who was considered to have an unusual body with large buttocks and enlarged genitalia, came from a slave family and was brought into Great Britain in 1810 to be exhibited and objectified by the public. Her body did not fit the ideas of a "normal" body. In such ways, people of color have been objectified historically for looking "different." In another well-documented example, Phineas Taylor Barnum conducted several exhibitions named "What is it?" throughout the nineteenth century. For these shows, Barnum employed a variety of people, such as Hervey Leech, an actor from New York with unusually small legs in proportion to the rest of his body. Leech was possibly Caucasian in origin, but disguised his hands and stained his face to play a person of color acting like "monkey" characters (jumping, grunting, and eating raw meat).[12] William Henry Johnson, an African American person of short stature with a developmental disability also participated in Barnum's exhibitions.[13] Leech and Johnson were exhibited to depict the liminality and hybridity between man and animal. These people were usually exhibited as "nondescript" or people who could not be described, and were left for the audience to discuss openly. Onlookers often discussed their animal characteristics in relation to them being people of color.[14]

As such, bodies that are considered too short, too tall, too fat, or too thin are also rendered "abnormal" and therefore pathologized. Similar to Saartje, Leech, and Johnson, Charles Stratton (Barnum's "General Tom Thumb") and Lavinia Warren ("Little Queen") were diminutive people exhibited in Barnum's shows as human curiosities and worked for Barnum's shows as entertainers.[15] People are rendered disabled on the basis of either "atrophy or degeneration" or "hypertrophy or enlargement."[16] The able-bodied paradigm is based on the notion of "a whole body, a single and fixed set of physical abilities, while the disabled body is in some way incomplete."[17] These differences not only render specific bodies as different and therefore "abnormal," but they also push for one unique form of body in shape, appearance, as well as ability; thus, ways and means are employed to "discipline the body to conform to dictates of both gender and the ability system."[18] This is often done through the means of medicalization and pathologization of bodies. The stress of treatment and cure is not toward improving social, environmental, and economic infrastructures, but on individual bodies and getting them to conform to one standard norm.[19] This medicalization process helps in transforming the horror and fascination often associated with "monstrous" bodies into the scientific language of an illness that can then be rendered into a classificatory system as "normal" or having the "normal" as an ideal to be attained.[20]

Intersex bodies are often subjected to these normalizing procedures in order to "correct" them and get them to conform to being strictly male or female through medical management at a very early stage. The medical model accords the intersex body as pathological with claims that if corrective surgeries were not conducted on intersex people, they would lead their lives in misery.[21] However, there is no singular idea of being a male or a female, even among physicians. While some physicians may accord greater importance to the size of the external genitals, there are others who lay more significance on the results of gender tests and still others on the person's ability to reproduce. The ideas of normality are guided not just by medicine and biomedical discourses, but also by society's notions of what is normal or what is not. Thus, not only is normalcy constantly constructed, it is also constantly being produced.[22] In the context of the plethora of different opinions and notions of normality, "the medical presupposition that intersex characteristics are inherently disabling to social viability remains the taken-for-granted truth from which clinical practice proceeds";[23] despite the fact that many of these surgeries are known to cause irreversible damage to intersex people.[24]

With the shift from looking at individual bodies to that of the population by the state,[25] a process of "governmentality" was evolved in which bodies began to be classified under several markers, including race: gender, class, color, ability, and sexual orientation, among others. Bodies that adhered to the idea of the "normal" and the "able" were therefore more valued within this form of governmentality. This ability and normality of the body is also evaluated in terms of (re)productivity for the nation state. Therefore, bodies that are more productive (thus healthy and able bodies without any disease or illness) and can reproduce (healthy) children who can raise the nation's worth, are valued. More specifically in terms of reproduction, certain bodies are more valued (e.g., white, heterosexual, nondisabled bodies). Intersex people and people with disabilities, even if they can be (re)productive are not valued within this framework, as it is feared that they can either not reproduce; or if they did, they would produce more of their "own kind," thus proving a burden on the community and the nation state. Both intersex bodies as well as disabled bodies are therefore rendered "abnormal" within this framework and also subjected to processes of corrections and alterations to make them "more normal."

Genetic Engineering and Eugenics

Such efforts to mark the "perfect body" through the lens of sexuality, ability, and race have been evident historically. However, they appeared in one of their most brutal forms during World War II, in concentration camps and through the extermination of lives that were considered burdensome to the nation state and therefore not worth living (be that Jewish, homosexual, disabled, Roma, etc.). Although its emphasis on eugenics was diminished after World War II, genetic engineering continues even today in subdued and subtler forms. This becomes visible, for example, through the medical termination of pregnancies based on disability. Abortions based on disability are often built into the laws and policies addressing the medical termination of pregnancies. For example, in India, medical termination of pregnancies is allowed up to 20 weeks in cases where prenatal congenital defects are detected or where the fetus carries a substantial risk of a physical or mental disability. While these abortions are allowed within the first five months of pregnancy, some congenital defects can only be diagnosed by the twentieth week.[26] This was highlighted in the Niketa Mehta case in 2008 in Mumbai, India, where Niketa's plea for abortion of her fetus, which was diagnosed with a serious heart defect, was rejected by the Mumbai High Court because her pregnancy had advanced beyond 20 weeks.[27] While the abortion was denied in this case, the medical termination of Mehta's pregnancy on the basis of disability would have been legal had she approached the hospital within the stipulated time of 20 weeks. Although this case is an extreme example, it was instrumental in bringing to fore the debates on women's right of abortion and the right of people with disabilities to be born. Although in this case the fetus had little possibility of survival even after birth, the clause of "substantial risk of a physical or mental" disability remains open to the interpretation of medical professionals and the family and may be used to terminate viable fetuses too solely on the basis of their ability status. The debate around the right of abortion for the woman is a complex one and beyond the scope of this chapter. My aim here is merely to highlight the incidence of abortions on the basis of the disability of the fetus, the "substantial risk" of the disability remaining open to interpretation and thus also being open to possible (mis)use. The elimination of fetuses with congenital defects[28] is to ensure the birth of "normal" babies without any issues, so that they may participate effectively in the (re)productivity of the family and the state. Inherent in this logic is the notion that people with disabilities cannot participate productively for the nation state.

According to Garland-Thomson,[29] the "socio-medical project of eradicating disability all too often is enacted as a program to eliminate people with disabilities through such practices as forced sterilization, so-called physician-assisted suicide and mercy killing, selective abortion, institutionalization [sic], and segregation policies." On one hand, Garland-Thomson[30] talks about the abortion of fetuses based on disability as a form of genocide against the disabled, questioning the ethics behind them; on the other, she discusses the right of the woman to choose whether to retain her pregnancy. Apart from these prenatal processes, people with disabilities may also be exposed to processes of growth attenuation to keep them small and therefore more manageable for caregivers. For example, this occurred in the controversial case of a girl called Ashley from the United States who was born with severe physical and developmental disabilities. Ashley was made to undergo a number of medical procedures, such as growth attenuation, a hysterectomy, and a double mastectomy, that questions the acceptable limits of medical intervention.[31] These procedures were undertaken with the view that they would make it easier for the parents and caregivers of patients with "profound" developmental disabilities (e.g., children who need help eating, using the toilet, and who are noncommunicative) to provide care and to involve them in family and social activities. Ashley could not sit up, walk, or talk. She could have grown physically as an adult and could have gone through regular biological processes of menstruation. However, she was subjected to a number of procedures in her prepubertal years that restricted her physical growth to a child of nine years.[32] These procedures can be justified from the perspective of the caregivers, who may find it difficult to care for disabled patients as they grow older and thus heavier. In addition, because of structural and systemic limitations, caregivers find little support from the state or the community in the process of caregiving, making it taxing and difficult. However, it is still significant to notice that in this initiative, where priority is accorded to the needs of the parents and caregivers, little attention is paid to the right of the person with disability to consent and to his/her bodily integrity, reaffirming his/her position as a partial citizen or noncitizen.

For intersex people, parents and physicians are rarely able to predict their intersexuality before birth.[33] However, with the advancement of science it has become increasingly possible to predict anomalies at the chromosomal- and genetic-level before birth. In the case of intersex people where it is not as easy to diagnose the intersexuality of the child before birth, this advancement in science and technology takes other forms. For example, in a recent development in India, there were reports of genital surgeries being conducted on children after birth to convert them from females to males.[34] Hardly any surgeries are conducted to convert males into females, although many doctors in the past have suggested that "it is easier to dig a hole than build a pole."[35] The news reports however did not mention any such surgeries on intersex infants. The preference for perfectly aligned sex, and particularly the male sex, seems to be gaining more ground in countries such as India. Although speculative at this juncture, and although the news article of conducting genital surgeries to convert females into males has been refuted (because of lack of evidence), it may not be long before science makes sufficient progress to be able to predict if the fetus is intersex. Although speculative, it is possible that intersex people may then be subjected to a fate similar to that of people with disabilities.

My aim here is not to argue against processes of genetic engineering, but rather to evaluate the motive behind these strategies. Advancement in technologies of medicine and health has also helped achieve better conditions of living. In this way, technological enhancements of the human body, termed by G. Berthoud as "techno-utopia," rely on the notion of the body as a biological organism that can be improved upon and on the belief that this kind of improvement is ethically correct.[36] Medical advancement has not just helped human beings to live longer and achieve better health standards, but it has also helped people to relieve pain and distress in bodies. The shift from the molar level to the molecular level in biopolitics, as instantiated by Rose,[37] has also become inevitable in certain ways. Thus, "molecularization is conferring a new mobility on the elements of life, enabling them to enter new circuits—organic, interpersonal, geographical, and financial."[38] This focus in looking beyond the molar level of organs and limbs to looking at the molecular level also comes with benefits not just for people in general, but also for intersex people and people with disabilities.

Miller argues that the womb is a paradigmatic space for biopolitics; therefore (re)productive politics play a major role in the attainment of rights as citizens.[39] Reproductive technologies play a vital role in determining who fits as a normal citizen and who does not. Those who possess wombs that produce normal, healthy, and nondisabled children are higher up on the hierarchy. With the advancement of reproductive technologies, the womb becomes separable from the body of woman (e.g., through test-tube babies); "the elements of reproduction—eggs, sperm, and later embryos—also become separable from any particular body, mobilized around circuits of laboratories, clinics, and other bodies."[40] Although this may mean that women stand to lose in not being the sole possessor of the womb, this could also signify the attainment in power of people who may generally not be considered within these power circuits. For example, these methods could be of help to infertile couples including some intersex people and people with disabilities to participate in reproductive processes and therefore be more assimilable within society.

Therefore, there are clear advantages to these kinds of medical and scientific procedures including molecular technology and genetic engineering. However, there is a need to examine and question the aim for these procedures and what they might entail. For example, we need to examine why it may be considered appropriate for an infertile couple to seek an egg donor with particular characteristics of height, weight, and intelligence,[41] but not so for a lesbian couple who were hearing impaired to seek a sperm donor who had a history of deafness within the family so that they may have a deaf child.[42] Public condemnation toward opting for a deaf child may come from a place where, at a systemic and structural level, people with disabilities and their caregivers have a more difficult time adjusting within society. However, is it not more important to challenge these systems and structures that inhibit wider access and resources to people who may not be considered normal, rather than to prevent the birth of children with different needs altogether? What determines boundaries of normality, who decides, and on what basis? Why does it become so important that people with all their diversities have to be assimilated within narrow standards of normality? Standards of normalization have therefore been used to benefit some people who adhere to these norms while marginalizing others. Consequently, it is supremely important that we examine these ideas of normality within biopolitics and explore how they impact people's lives. Molecularization cannot stand alone; it must be combined with standardization, regulation, and ethics, and as such, "at this molecular level...life itself has become open to politics."[43]

The Exceptions

Through the process of governmentality, as discussed by Foucault, the extension of power by the state onto physical and political bodies took two forms: one with the body as "a machine," which can be disciplined and which could work to its optimum capacities; and the other as "species body" or the body that formed the basis of biological processes such as reproduction, birth, mortality, and level of health.[44] Thus under this process of governmentality, bodies are compartmentalized in different ways: able/disabled, (re)productive or not. Bodies are therefore neatly categorized, and all those who may not fit in within the able-bodied and (re)productive paradigm fall through the cracks. Both, people with physical disabilities as well as intersex people are considered to have bodies that do not fit notions of normality. Parallels can be drawn between intersex people and people with disabilities especially in the arena of sports. People with disabilities are often not considered able and fit to participate in sports activities along with able-bodied people. Representations of people with disabilities portray them as weak, lacking in ability and physical strength. They are also often infantilized. Although the first sports competition for disabled people was held in 1948 for World War II veterans with spinal cord injuries, it was in 1960 that the first Paralympics games were held in Rome with Olympic-style games. Over the years, there has been an increase not just in the number of athletes participating in the Paralympics games, but also in the variety of events organized for people with disabilities.[45]

Separate sports events for people with disabilities suggests that disabled persons are not considered at par with nondisabled people, as they require a separate set of parameters for being judged. Using the same parameters for people with disabilities as used for nondisabled people was considered to put the former at a disadvantage. However, the juxtaposition of ability versus disability takes a different turn when considering sportspeople such as Oscar Pistorius and Aimee Mullins. Oscar Pistorius, also popularly referred to as "the blade runner," is a double amputee athlete who uses the Cheetah artificial limbs. He participated in sports events for nondisabled persons for the first time in 2007. His participation in sports for nondisabled people, however, generated debates about him having an unfair advantage over the other players because of his artificial limbs. Aimee Mullins is an actress, athlete, and fashion model, as well as a double amputee. Similar to Pistorius, Mullins also is known for competing with nondisabled people during her college years at Georgetown University. Accomplishments such as those of Pistorius and Mullins instantly raise questions such as what constitutes ability and how much does technology and human perseverance have to do with one's ability. People who in general are considered to be at a disadvantage to others because their bodies do not fit the stereotype of "normal" are suddenly considered to pose a threat to other athletes.[46]

Similar debates are raised for intersex people within the sports arena. Athletes such as Caster Semenya and Santhi Soundarajan are considered to have an unfair advantage over other women athletes. Caster Semenya, an athlete from South Africa, won the gold medal in the 800 meters at the 2009 World Championships. Although she had won the World Junior Championships the previous year as well, her World Championship time improved by about eight seconds, setting a world record.[47] Her accomplishment raised suspicion among her competitors. Even though she was not suspected of cheating, she was assumed to possess an unfair advantage because her rivals believed that, despite competing as a woman, she was male.[48] Semenya was muscular and had some facial hair, which may have prompted the misconception; additionally, it could have been because of excessive androgen, which is considered a male hormone.[49] Irrespective of the test results, which were inconclusive, Semenya was banned from the games for almost 11 months and was finally allowed to compete in July 2010 against other female athletes.[50]

In a similar case, Santhi Soundarajan, an Indian athlete, was stripped of her silver medal in the 2006 Asian games after she failed to pass the gender tests. Reports in the media speculate about her having an intersex "condition" called the Androgen Insensitivity Syndrome (AIS) with general physical characteristics of a female but with a genetic makeup that includes a male chromosome.[51] During Semenya's case, the International Olympic Committee had met to formulate guidelines for cases of indeterminate gender. However, these guidelines were based on certain misconceptions, including that there is a clear way to distinguish between sexes, that intersexuality as a "condition" can be spotted visually, or that the intersex person needs "correction" such as hormone therapy, even at the cost of the person's basic health.[52]

"Abnormals" or "Exceptions" • 53

The intersection between these two sets of cases—people with disabilities and people who do not conform to being male or female—is the normalization regime around bodies that deem certain bodies as "normal by nature" whether it is in terms of one's ability or one's sex. In the field of sports, bodies are compartmentalized around different norms. These compartmentalizations work toward creating neat boundaries, whether it is around those of sex or those of ability. People are required to compete only with people of the same sex so as not to be at a disadvantage or pose an unfair advantage over the others. Similarly, people can compete only with others who have similar abilities in order for the competition to be fair. These compartmentalizations often happen in binaries of male/female or disabled/nondisabled, without taking into consideration that there may not be neat divisions across gender or ability. Because the occurrence and the experience of disability are not uniform, laws and policies in different countries use different parameters in defining disabilities. If the lines of distinction between the nondisabled and disabled could be drawn so neatly, perhaps these laws would not differ. Similarly, sex continues to be considered in binary terms although the points of distinction between males and females are not so clear. Unlike in early gender tests, when the lines of distinction were drawn around the external genitalia of athletes, in contemporary times the distinctions become far less conclusive. They are not based on the external genitalia alone, but instead involve a series of gender tests that are conducted on the participants. This blurriness indicates how societal and cultural factors influence notions of sex.

My aim here is not to argue for people with disabilities to be mainstreamed with the nondisabled or to argue that intersex people be included within the category of "females." I argue instead for questioning the normalization procedures that compartmentalize people into neat binary categories of sex or ability. I also argue for analyzing the administrative procedures that necessitate making these neat boundaries in different areas, including sports. Why does it seem so important to have these fine lines of distinction on the basis of sex or ability? Is it because certain sexes or certain bodies seem more "able" than others? It is interesting to note here that intersex bodies that are in general construed as "abnormal" are considered more at an advantage to those categorized as "female," at least in the field of sports.

Ideas about normalization however vary across different parameters. For example, intersex bodies are construed as "not so able" in matters such as reproduction. Bodies therefore get categorized on the basis of certain functions, abilities, size, looks, and so on. Whereas people with disabilities can have their own sports events, they become more of a threat when they compete and have a perceived winning edge over non-disabled individuals; suddenly the disabled bodies cease to be weak and incapable, due to the use of technology and artificial limbs. Similarly, there has not been an equivalent debate on whether intersex people can compete with males. This is partly because they are not yet considered a threat to men and their abilities. Although they may be considered stronger than females, they are considered not strong enough as males. Would it matter only if they begin to have a winning edge over the males in the sporting events?

Exceeding the biological limitations of the body also means exceeding the human body itself; in other words in a utopian project, freeing the human body from pain and disease can be achieved with the separation of human existence from the body, thus leading to a transformation of the person into a virtual entity.[53] On one hand, utopia could be a place where one can have a body without a body (i.e., the utopia of an incorporeal body that is infinite in its capacities), while on the other hand, utopia could also be a place for erasing bodies (i.e., a place where the body is negated and transfigured).[54] However, in working toward an understanding of the disabled body, the intersex body, or other "monstrous bodies," it is important to recognize the impossibility of a "perfect" body that is stable; bodies are always fluid, vulnerable, ever-changing.[55] This calls for a transgressive utopianism that does not use utopia as an inscription of perfection, which usually arises from a place of discontent and dissatisfaction, but rather offers an alternative perspective that is creative and has a subversive or transformative potential.[56]

Conclusion

These categorizations and compartmentalizations of bodies can be evaluated on the process of governmentality, which considers some bodies as more "able" and therefore more productive. Although this grid of normality when applied to the whole population should work for most people, it is not geared to address people and situations where it does not work. As an administrative process therefore it does not do justice to people and their realities and experiences. According to this process, the bodies that do not fit (whether in terms of their sex or their ability) are construed as "abnormals," which in turn influences how the "abnormals" are treated. While their bodies may function as any other body and they may be productive in a variety of ways, intersex and disabled bodies are still not seen as equivalent to normal bodies and are thus subjected to medical interventions. They become construed as weak, unproductive, and unworthy of reproductive function. However, in the cases where some people with disabilities and intersex people are shown to excel with the use of technology, such as sports, intersex and disabled bodies start to pose a threat to the nondisabled. As a result, fresh boundaries are assigned on the acceptable limits of normalcy. In these exceptional circumstances, bodies that surpass the boundaries of normalcy are still not considered "normal" and therefore do not enjoy similar rights. It is thus vital that the boundaries of normality are constantly evaluated; these ever-shifting parameters of normality must be regularly examined to work toward inclusion and diversity within our work, our abilities, our choices, and our desires.

Notes

1. Jarosław Barański, "The Plastic Body as an Epistemological Figure of Biotechnological Utopia," *Studia Philosophiae Christianae* 47 (2011): 5–13.
2. Rosemarie Garland-Thomson, "Integrating Disability, Transforming Feminist Theory," *NWSA Journal* 14 (2002); Minae Inahara, "This Body Which Is Not One: The Body, Femininity and Disability," *Body & Society* 15 (2009); Sophie Mitra, "The Capability Approach and Disability," *Journal of Disability Policy Studies* 16 (2006); and Margrit Shildrick, *Embodying the Monster: Encounters with the Vulnerable Self* (London: Sage, 2002).
3. Garland-Thomson.
4. Michel Foucault, "From the Power of Sovereignty to Power Over Life, Lecture (17 March 1976)," in *"Society Must be Defended": Lectures at the Collège de France, 1975–76,* edited by Mauro Bertani et al. (New York: Picador, 2003).
5. Nikolas Rose, *The Politics of Life Itself: Biomedicine, Power, and Subjectivity in the Twenty-first Century* (Princeton: Princeton University Press, 2007).
6. Foucault," Powero fS overeignty."
7. Ibid.
8. Rose, *The Politics of Life Itself.*
9. Giorgio Agamben, *Homo Sacer: Sovereign Power and Bare Life,* trans. Daniel Heller-Roazen (Stanford: Stanford University Press, 1998).
10. "What is Intersex?: Defining Intersex," OII-Australia (Organization Intersex International Australia), last modified August 2, 2013, http://oii.org.au/18106/what-is-intersex/.
11. Garland-Thomson, "Integrating Disability."

12. James W. Cook, "Of Men, Missing Links, and Nondescripts: The Strange Career of P. T. Barnum's 'What Is It?' Exhibition," in *Freakery: Cultural Spectacles of the Extraordinary Body*, edited by Rosemarie Garland-Thomson (New York: New York University Press, 1996).
13. Ibid.
14. Ibid.
15. Lori Merish, "Cuteness and Commodity Aesthetics: Tom Thumb and Shirley Temple," in *Freakery: Cultural Spectacles of the Extraordinary Body*, edited by Rosemarie Garland-Thomson (New York: New York University Press, 1996).
16. Garland-Thomson, "Integrating Disability," 7.
17. Inahara, "This Body Which Is Not One," 59.
18. Garland-Thomson, "Integrating Disability," 10.
19. Ibid.
20. Elizabeth Grosz, "Intolerable Ambiguity: Freaks as/at the Limit," in *Freakery: Cultural Spectacles of the Extraordinary Body*, edited by Rosemarie Garland-Thomson (New York: New York University Press, 1996).
21. Sharon E. Preves, "Sexing the Intersexed: An Analysis of Sociocultural Responses to Intersexuality," *Signs* 27 (2002): 523–556.
22. Ibid.
23. Morgan Holmes, "Rethinking the Meaning and Management of Intersexuality," *Sexualities* 5 (2002): 169.
24. Preves, "Sexing the Intersexed."
25. Foucault, "Power of Sovereignty."
26. Neha Madhiwalla, "The Niketa Mehta Case: Does the Right to Abortion Threaten Disability Rights?" *Indian Journal of Medical Ethics* 5 (2008): 152–153.
27. Ibid.
28. Apart from diagnosis of congenital defects in fetuses, the media is also rife with reports on the discovery of the "gay gene." According to some studies, homosexuality can be traced to genes (Steve Connor, "The 'Gay Gene' is Back on the Scene," *The Independent,* November 1, 1995, http://www.independent.co.uk/news/the-gay-gene-is-back-on-the-scene-1536770.html). However, it is relevant to consider why it is important to find out whether homosexuality is due to genetic factors. Research on the "gay gene" by Dean Hamer found support from a number of people who were interested in not just the investigation of such a gene, but also the possibility of eradication through prenatal tests and the abortion of fetuses that may exhibit particular genetic markers (Ibid.). Discourses around physical disability and sexuality perpetuate the notion of "normal" bodies, thus marking bodies that do not conform to the heteronormative order as "abnormal." As a result, the search for the gay gene aims toward rearing the perfect, heterosexually reproductive couple, so they in turn can reproduce healthy, heterosexual, (and by that logic) "normal" babies. Notions of abnormality are not restricted to discourses of disability alone (although these practices of elimination happen too often on disabled bodies), but rather they center on what is considered normal in contemporary society at any period of time.

29. Garland-Thomson, "Integrating Disability," 15
30. Ibid.
31. Emi Koyama, "Growth Attenuation Treatment Going Mainstream, and the Limits of Disability Studies," *Eminism*, January 27, 2009, http://eminism.org/blog/entry/48.
32. Ed Pilkington, "Frozen in Time: The Disabled Nine-year-old Girl Who Will Remain a Child All Her Life," *The Guardian*, January 4, 2007, http://www.guardian.co.uk/world/2007/jan/04/health.topstories3.
33. Joel Frader et al., "Health Care Professionals and Intersex Conditions," *Archives of Pediatrics & Adolescent Medicine* 158 (2004): 426–429.
34. Amrita U. Kadam, "Docs Turn Baby Girls into Boys," *Hindustan Times*, June 26, 2011, http://www.hindustantimes.com/News-Feed/madhyapradesh/Indore-doctors-turn-scores-of-baby-girls-into-boys/Article1-713863.aspx.
35. Preves, "Sexing the Intersexed"; Holmes, "Rethinking the Meaning and Management of Intersexuality," 169.
36. Barański, "The Plastic Body as an Epistemological Figure of Biotechnological Utopia."
37. Rose, *The Politics of Life Itself*.
38. Ibid., 15.
39. Ruth A. Miller, *The Limits of Bodily Integrity: Abortion, Adultery, and Rape Legislation in Comparative Perspective* (Hampshire: Ashgate, 2007).
40. Rose, *The Politics of Life Itself*, 14.
41. Michael J. Sandel discusses a case where an infertile couple sought an egg donor "who was 5'10 tall, athletic, without major family medical problems and to have a combined SAT score of 1,400 or above." The couple offered a payment of $50,000 to such egg donor (*The Case Against Perfection: Ethics in the Age of Genetic Engineering* [Cambridge: Belknap Press, 2007], 1–2.).
42. Sandel also discusses the case of Sharon Duchesneau and Candy McCullough who preferred to have a deaf child by seeking a sperm donor "with five generations of deafness in family" (Ibid.). Sandel also discusses how this story published in the *Washington Post* received extensive condemnation from the public.
43. Rose, *The Politics of Life Itself*, 15.
44. Foucault, "Power of Sovereignty," 139.
45. IPC (The International Paralympic Committee), "Paralympic Games," accessed March 13, 2014, http://www.paralympic.org/Paralympic_Games/.
46. Aimee Mullins, "Racing on Carbon Fiber Legs: How Abled Should We Be?" *Gizmodo*, November 12, 2009, http://gizmodo.com/5403322/racing-on-carbon-fiber-legs-how-abled-should-we-be.
47. BBC News, "Birth Certificate Backs SA Gender," *BBC News*, August 21, 2009, http://news.bbc.co.uk/2/hi/africa/8215112.stm.
48. Sarah Leonard, "Sex, Sport, and Scandal," *Queen's Medical Review* 3 (2010): 6–7.
49. Ibid.

50. Anna Kessel, "Caster Semenya May Return to Track This Month after IAAF Clearance," *The Guardian,* July 6, 2010, http://www.guardian.co.uk/sport/2010/jul/06/caster-semenya-iaaf-clearance.
51. Emine Saner, "The Gender Trap," *The Guardian,* July 30, 2008, http://www.guardian.co.uk/sport/2008/jul/30/olympicgames2008.gender.
52. In addition, there are also cases where athletes have been administered steroids at competitive sports events so that they may outdo other performers. For example, Dr. Robert Kerr was reported to have used anabolic steroids on almost 20 medalists in the 1984 Olympic Games; Pat Connolly, a coach of the women's track team has also reported that an estimated five out of ten medalists had been administered steroids in the Seoul Olympics (John O'Leary, *Drugs and Doping in Sport: Sociolegal Perspectives* [London: Cavendish, 2001]). These cases are pertinent because they illustrate how technology can help in creating bodies desirable for events such as competitive sports, though they are considered "abnormal" otherwise. Many of the sportspeople who had received such drugs have reported a number of health-related issues as a result. For example, female athletes who have been administered such drugs reported experiencing hormonal disturbances, liver and kidney problems, and developing male characteristics such as developing excessive body hair, muscles, and deep voices (Ibid.). The case of Andreas Krieger, a female to male transgender man who was a popular short putter from the German Democratic Republic, brings together the issues of steroid use among athletes and gender identity. Krieger participated in and won many international sports competitions, including getting the gold medal in the 1986 European Championship in Athletics. He was also administered anabolic steroids (José Luis Pérez Triviño, *The Challenges of Modern Sport to Ethics: From Doping to Cyborgs* [Lanham: Lexington Books, 2013]). Manfred Ewald, the former head of the East German Sports Association, and his former medical director, Dr. Manfred Hoeppner, were both found guilty of administering such drugs to almost 142 East German sportswomen and were sentenced to prison (O'Leary). Krieger was one of the athletes who testified against Ewald and Hoeppner, and he mentioned that the steroids had left him with masculine characteristics and had had a deep impact on his health (Triviño). He made the decision to undergo gender reassignment surgery later in his life, according the cause of his "transsexuality" to the administration of the steroids (Ibid.). Although this case illustrates an extreme, where the administration of drugs to alter female athletes for a competitive advantage led to the person opting for a gender-reassignment surgery, it is an important example of how technology has been and continues to be used to alter bodies. Moreover, these alterations are not just to enhance performances in competitive sports, making them desirable in certain contexts; they may also render the body undesirable in other contexts, proving detrimental to the individual's health. Elizabeth Reis, "Is Intersex a Disorder or a Competitive Advantage?" Accessed June 1, 2014, http://www.womensbioethics.org/index.php?p=Is_Intersex_a_Disorder_or_a_Competitive_Advantage&s=355.

53. Barański "The Plastic Body as an Epistemological Figure of Biotechnological Utopia."
54. Michel Foucault, "Utopian Body," in *Sensorium: Embodied Experience, Technology, and Contemporary Art,* edited by Caroline A. Jones (Cambridge: MIT Press, 2006).
55. Inahara, "This Body Which Is Not One."
56. Lucy Sargisson, *Utopian Bodies and the Politics of Transgression* (New York: Routledge, 2000).

Bibliography

Agamben, Giorgio. *Homo Sacer: Sovereign Power and Bare Life,* translated by Daniel Heller-Roazen. Stanford: Stanford University Press, 1998.
Barański, Jarosław. "The Plastic Body as an Epistemological Figure of Biotechnological Utopia." *Studia Philosophiae Christianae* 47 (2011): 5–13.
BBC News. "Birth Certificate Backs SA Gender." *BBC News,* August 21, 2009. Accessed April 2, 2014. http://news.bbc.co.uk/2/hi/africa/8215112.stm.
Connor, Steve. "The 'Gay Gene' is Back on the Scene." *The Independent,* November 1, 1995. Accessed January 2, 2013. http://www.independent.co.uk/news/the-gay-gene-is-back-on-the-scene-1536770.html.
Cook, James W. "Of Men, Missing Links, and Nondescripts: The Strange Career of P. T. Barnum's 'What Is It?' Exhibition." In *Freakery: Cultural Spectacles of the Extraordinary Body,* edited by Rosemarie Garland-Thomson, 139–157. New York: New York University Press, 1996.
Foucault, Michel. "From the Power of Sovereignty to Power Over Life, Lecture (17 March 1976)." In *"Society Must be Defended": Lectures at the Collège de France, 1975–76,* edited by Mauro Bertani and Alessandro Fontana, 239–264. New York: Picador, 2003.
———. "Utopian Body." In *Sensorium: Embodied Experience, Technology, and Contemporary Art,* edited by Caroline A. Jones, 229–234. Cambridge, USA: MIT Press, 2006.
Frader, Joel, Priscilla Alderson, Adrienne Asch, Cassandra Aspinall, Dena Davis, Alice Dreger, James Edwards, Ellen K. Feder, Arthur Frank, Lisa A. Hedley, Eva Kittay, Jeffrey Marsh, Paul S. Miller, Wendy Mouradian, Hilde Nelson, and Erik Parens. "Health Care Professionals and Intersex Conditions." *Archives of Pediatrics & Adolescent Medicine* 158 (2004): 426–429.
Garland-Thomson, Rosemarie. "Integrating Disability, Transforming Feminist Theory." *NWSA Journal* 14 (2002): 1–32.
Grosz, Elizabeth. "Intolerable Ambiguity: Freaks as/at the Limit." In *Freakery: Cultural Spectacles of the Extraordinary Body,* edited by Rosemarie Garland-Thomson, 55–66. New York: New York University Press, 1996.
Holmes, Morgan (2002). "Rethinking the Meaning and Management of Intersexuality." *Sexualities* 5 (2002): 159–180.

Inahara, Minae. "This Body Which Is Not One: The Body, Femininity and Disability." *Body & Society* 15 (2009): 47–62.
Kadam, Amrita U. "Docs Turn Baby Girls into Boys." *Hindustan Times*, June 26, 2011. Accessed March 30, 2014. http://www.hindustantimes.com/News-Feed/madhyapradesh/Indore-doctors-turn-scores-of-baby-girls-into-boys/Article1-713863.aspx.
Kessel, Anna. "Caster Semenya May Return to Track This Month after IAAF Clearance." *The Guardian*, July 6, 2010. Accessed June 5, 2014. http://www.guardian.co.uk/sport/2010/jul/06/caster-semenya-iaaf-clearance.
Koyama, Emi. "Growth Attenuation Treatment Going Mainstream, and the Limits of Disability Studies." *Eminism*, January 27, 2009. http://eminism.org/blog/entry/48.
Leonard, Sarah. "Sex, Sport, and Scandal." *Queen's Medical Review* 3 (2010): 6–7.
Madhiwalla, Neha. "The Niketa Mehta Case: Does the Right to Abortion Threaten Disability Rights?" *Indian Journal of Medical Ethics* 5 (2008): 152–153.
Merish, Lori. "Cuteness and Commodity Aesthetics: Tom Thumb and Shirley Temple." In *Freakery: Cultural Spectacles of the Extraordinary Body*, edited by Rosemarie Garland-Thomson, 185–203. New York: New York University Press, 1996.
Miller, Ruth A. *The Limits of Bodily Integrity: Abortion, Adultery, and Rape Legislation in Comparative Perspective*. Hampshire, UK: Ashgate, 2007.
Mitra, Sophie. "The Capability Approach and Disability." *Journal of Disability Policy Studies* 16 (2006): 236–247.
Mullins, Aimee. "Racing on Carbon Fiber Legs: How Abled Should We Be?" *Gizmodo*, November 12, 2009. http://gizmodo.com/5403322/racing-on-carbon-fiber-legs-how-abled-should-we-be.
OII-Australia (Organization Intersex International Australia) "What Is Intersex?: Defining Intersex." Last modified August 2, 2013. Accessed May 28, 2014. http://oii.org.au/18106/what-is-intersex/.
O'Leary, John. *Drugs and Doping in Sport: Socio-Legal Perspectives*. London: Cavendish, 2001.
Pilkington, Ed. "Frozen in Time: The Disabled Nine-year-old Girl Who Will Remain a Child All Her Life." *The Guardian,* January 4, 2007. Accessed May 9, 2014. http://www.guardian.co.uk/world/2007/jan/04/health.topstories3.
Preves, Sharon E. "Sexing the Intersexed: An Analysis of Sociocultural Responses to Intersexuality." *Signs*, 27 (2002): 523–556.
Reis, Elizabeth. "Is Intersex a Disorder or a Competitive Advantage?" Accessed June 1, 2014. http://www.womensbioethics.org/index.php?p=Is_Intersex_a_Disorder_or_a_Competitive_Advantage&s=355.
Rose, Nikolas. *The Politics of Life Itself: Biomedicine, Power, and Subjectivity in the Twenty-first Century*. Princeton: Princeton University Press, 2007.
Sandel, Michael J. *The Case Against Perfection: Ethics in the Age of Genetic Engineering*. Cambridge, USA: Belknap Press, 2007.

Saner, Emine. "The Gender Trap." *The Guardian*, July 30, 2008. Accessed February 24, 2014. http://www.guardian.co.uk/sport/2008/jul/30/olympicgames2008.gender.

Sargisson, Lucy. *Utopian Bodies and the Politics of Transgression*. New York: Routledge, 2000.

Shildrick, Margrit. *Embodying the Monster: Encounters with the Vulnerable Self*. London: Sage, 2002.

IPC (The International Paralympic Committee). "Paralympic Games." Accessed March 13, 2014. http://www.paralympic.org/Paralympic_Games/.

Triviño, José Luis Pérez. *The Challenges of Modern Sport to Ethics: From Doping to Cyborgs*. Lanham: Lexington Books, 2013.

CHAPTER 3

The Inauspicious Regulatory Beginnings of Preimplantation Genetic Diagnosis

Patricia Stapleton

With recent advancements in the field of genetic engineering and its biotechnology applications, scientists have been able to develop new technologies to assist in human reproduction. While seemingly presented as a medical utopia (the answer to infertility, the end to genetic diseases, etc.), assisted reproductive technologies (ART) also present serious ethical and regulatory challenges. The lack of a strong regulatory framework in the United States raises immediate ethical concerns about ART: "family balancing" through sex-selective abortions, eugenic practices, and the implications of outsourcing surrogacy to developing countries, to name a few. In addition, these ethical challenges fall squarely into the realm of biopolitical debates. Not only does the state exert power over the life and bodies of its citizens in other spheres, the government's regulation of reproduction (or the lack thereof) even influences who its future citizens will be, with potentially dystopian results.

This chapter engages with both the utopian drive to "perfect" pregnancy and childbirth in the United States and the possible dystopian outcomes from loosely regulated ART. To do so, it reviews current ART practices in the United States, with a particular focus on the preimplantation genetic diagnosis (PGD) procedure. It then addresses the existing American regulatory framework and its weaknesses. The chapter also provides an outline of the ethical and regulatory challenges that continued developments in ART pose for the American government, medical community, and women. Finally, it explores how the development of ART can be born out of a utopian desire to perfect pregnancy, child birth, and humans themselves; and how the government's current regulatory position could lead to dystopian results.

Theoretical Concepts

Before delving into a review of the availability and usage of ART in the United States, we must first review how the concepts of "utopia" and the "body" are employed in this work. Both concepts are used with varying definitions across disciplines, and scholars may engage with these ideas in vastly different ways. In this chapter, "utopia" as a concept is rooted firmly in Ruth Levitas's work, while "body" is drawn from Bryan S. Turner's research. Thus, we rely on a more sociological understanding of both concepts to inform our study of the lack of regulation of PGD.

In *The Concept of Utopia*, Levitas seeks to clarify the meaning of utopia and offer a new definition that accounts for the use of the concept across disciplines.[1] She notes that utopia is often described by three different elements: content, form, and function. In this chapter, the focus is on function, where utopia presents "some kind of goal, even if commentators as opposed to the authors of utopias do not see them as necessarily realisable in all their details."[2] The utopian drive within new technologies for reproductive health—particularly those that rely on genetic testing—implies that human bodies can be improved upon, and that we can make these improvements even before birth through informed selection of particular traits. In this way, medical professionals act as the "authors" of a utopia where no one would suffer from debilitating ailments that could be avoided through ART.

As Levitas states, however, utopias that center on function and the possibility of progress "raise questions about what the goal should be."[3] The utopian motivations that drive innovation in ART, like the use of PGD to prevent the birth of children with debilitating conditions, are founded on the belief that patients will use these techniques to create a better world—a world with less suffering. Yet, the use of genetic testing in ART, discussed in more detail below, raises profound ethical and moral questions about its accessibility and use. "Commentators" also challenge the notion that eliminating "undesirable" characteristics from the gene pool through PGD will have utopian results. Consequently, the lack of regulation of ART in the United States allows patients with the means and opportunity to pursue individual interests in their reproductive choices, and while that may be the elimination of a genetic disorder that would cause severe harm to their children, it can also mean embryo selection based on more superficial characteristics, like gender. In this way, the body of a child is perceived as a potentiality: the possibility of what could be and a physical site of prospective progress.

Turner's theoretical work thus helps us explore how the utopian drive underlying ART views the "body as a potentiality which is elaborated by culture and developed in social relations."[4] The body, of course, is more than just its malleability; Turner argues that the body also can be defined as an organism, a system of representation, and a lived experience.[5] It is all of these things together, as these descriptions of the human body are interrelated. But when we look at the utopian aspirations of ART and its usage, the body is most clearly viewed as a site for improvement, for example the potential parent whose infertility can be overcome or the potential child whose genotypic characteristics can be carefully selected with medical intervention.

Hence, a medical approach is reductive in its treatment of the body and its potential. If we return briefly to Levitas, she posits that "sometimes utopia embodies more than an image of what the good life would be and becomes a claim about what it could and should be: the wish that things might be otherwise becomes a conviction that it does not have to be like this. Utopia is then not just a dream to be enjoyed, but a vision to be pursued."[6] By joining this perspective on utopia with Turner's concept of the body as potentiality, ART becomes an instrument through which the conviction that we can improve our bodies can be pursued. Humans do not have to suffer; ART can conquer infertility and genetic disorders through scientific progress. But this reductionist approach to reproductive health raises questions about what the goals should be for these convictions. In expanding the definition of the body to include all of Turner's elements, we can more appropriately review the social context of ART and possibly answer those questions. Furthermore, we will be able to see how debate over a regulatory framework brings social, cultural, and political context back into our discussion of ART's utopian project.

Assisted Reproductive Technologies and PGD

Like the underlying concepts of utopia and the body, definitions of ART can vary. Here, we use the definition provided by the Centers for Disease Control and Prevention (CDC). The CDC's definition is based on the 1992 Fertility Clinic Success Rate and Certification Act that requires the CDC to publish the annual ART Success Rates Report:

> ART includes all fertility treatments in which both eggs and sperm are handled. In general, ART procedures involve surgically removing eggs from a woman's ovaries, combining them with sperm in the laboratory, and returning them to the woman's body or donating them to another woman. They do NOT include treatments in which only sperm are handled (i.e., intrauterine—or artificial—insemination) or procedures in which a woman takes medicine only to stimulate egg production without the intention of having eggs retrieved.[7]

The use of ART in the United States has doubled over the past decade, although its use is "still relatively rare when compared to potential demand."[8] Even so, in the last few years the CDC has noted increases in its ART Fertility Clinic Success Rates Reports across the categories of ART cycles, live births, and live infants (see Table 3.1).[9]

In a span of just three years, the number of ART cycles performed increased by 28,015 cycles (18.4 percent). About 32 percent of ART cycles performed in 2010 led to a live birth, defined as the delivery of one or more living infants; that number fell to approximately 29 percent in 2012. While the incidence of multiples (twins, triplets, etc.) fell between 2010 and 2012, over 28,000 more ART cycles were performed for only 4,204 additional live births. Although doctors have been successful at reducing the rate of births of multiples (lowering the risks of complications associated with them), increased use of ART has not translated into better success rates for patients. Despite this decline in clinic success rates, "over 1% of all infants born in the United States every year are conceived using ART."[10]

Table 3.1 ART fertility clinic success rates reports for 2010, 2011, and 2012

Year	2010	2011	2012[a]
# of ART cycles performed	147,260	151,923	176,275
# of reporting clinics	443	451	456
# of live births	47,090	47,818	51,294
#of live infants	61,564	61,610	65,179

[a]The 2012 report is preliminary.

Not all assisted reproductive technologies use biotechnology methods. But within the range of available ART procedures, the PGD process does. The CDC defines PGD as:

> A technique combining the recent significant advances in molecular genetics and ART. PGD allows physicians to identify various genetic diseases in the embryo (fertilized egg with several divisions) prior to implantation, that is, before the pregnancy is established. It is of special value for those who are at risk of having children with serious genetic problems.[11]

When selected, the PGD procedure is always used in conjunction with in vitro fertilization (IVF) to screen for specific genetic or chromosomal abnormalities before transferring the fertilized eggs into the mother. The technician removes one or two cells from the embryo about three to eight days after fertilization, then examines the chromosomes for obvious abnormalities such as trisomies (three chromosomes instead of the normal two) or aneuploidy (change in the number of chromosomes that can lead to a chromosomal disorder).[12] PGD is recognized as an important alternative to prenatal diagnosis (like amniocentesis). However, diagnosis from a single cell remains a technically challenging procedure, and the risk of misdiagnosis cannot be eliminated.[13]

PGD is used to test for a suite of genetic disorders, including cystic fibrosis, Tay-Sachs disease, muscular dystrophy, and hemophilia.[14] Recent developments have also led doctors to use PGD to select genetically matching embryos to aid in the treatment of sick siblings through the donation of cord blood or bone marrow.[15] Additionally, its use in diagnosing trisomy and aneuploidy helps doctors determine which embryos would be more suitable for implantation, as embryos with these chromosomal issues are less likely to become viable pregnancies.[16] Finally, PGD can determine the sex chromosomes of the embryo, which allows doctors to avoid implanting embryos that may carry specific gene markers on the X or Y chromosome. In general, it is also an accepted practice in the United States to select embryos for implantation based on sex for family balancing or parental preference for a certain gender. In short, PGD can tell parents the health status and gender of their developing embryos, giving them the information they need to decide which embryos to implant into the woman's womb.[17]

PGD was first used in the United States in 1989,[18] but comprehensive data on the use of PGD do not exist.[19] Estimates indicate that over

1,000 babies worldwide have been born following the use of PGD in the IVF process.[20] But, while the ART Success Rates Reports do include percentages of cycles that use PGD in their process, the CDC does not require that clinics report the number of births resulting from procedures that specifically use PGD. In addition, the CDC has collected general data on the use of PGD in IVF cycles only since 2004. Since the data has been collected, average annual usage rates across clinics have hovered around 4–5 percent.[21]

The average usage across all reporting clinics, however, does not accurately reflect the widely varying rates of PGD usage at specific clinics.[22] Data from the preliminary 2012 report show that 153 out of 456 clinics (approximately 33 percent) reported zero percent to less than 1 percent as their rate of PGD usage. The majority of clinics (244 or approximately 54 percent) reported PGD usage rates between 1 and 10 percent. Despite these low rates of PGD usage for the vast majority of reporting clinics, approximately 13 percent use PGD in more than one out of ten ART cycles. Four clinics (slightly less than 1 percent of all reporting clinics) reported 50 percent or higher, with one Las Vegas clinic reporting a rate of 86 percent usage of PGD, although the clinic performed only a total of 196 cycles (or 0.1 percent of all ART cycles performed in the United States for 2012).[23]

The four clinics reporting 50 percent or higher for PGD usage rates only represent a tiny sliver of ART cycles completed in the United States, with a total of 1,188 (or 0.67 percent of all ART cycles). Yet, when looking at absolute numbers, these four clinics performed approximately 725 IVF cycles using PGD. And, if we assume that the 2012 preliminary data will yield a PGD usage rate of about 4.5 percent across all reporting clinics (comparable to the average use reported the last several years), approximately 7,900 IVF cycles will have incorporated the PGD procedure out of the 176,275 ART cycles performed in the United States in 2012.[24] Thus, despite low average rates of usage across clinics, a significant number of fertility patients are choosing to use PGD during their treatment. Furthermore, if we assume a success rate for ART cycles using PGD consistent with the success rate average of all reporting clinics (29 percent), that would indicate that approximately 2,300 live births in the United States were the result of IVF cycles using PGD in 2012 alone.[25] This number is significantly higher than the "over 1,000 babies" worldwide that is often cited in the literature on PGD.

Comparatively, the European Society for Human Reproduction and Embryology (ESHRE) PGD Consortium has collected annual data on

PGD usage from its members since 1997, when it was first established. Clinics from all over the world have submitted data to the consortium in its first ten years. The ESHRE's focus is on Europe, with 56 European clinics reporting at some point in that time period, while only nine clinics from the United States have submitted data for collection.[26] Nevertheless, in a review of ten years of available data (1997–2007) collected from around the world, the PGD Consortium noted that 5,187 clinical pregnancies resulted in 5,135 newborns.[27] The disparity in these estimates from the cited literature further supports the importance of collecting data about PGD usage.

Although the process of PGD offers a range of possibilities for its use, there is a dearth of data to indicate exactly why it is used in clinics. A study published in 2008 surveyed fertility clinics, in part to determine the reasons why patients selected PGD as a procedure.[28] But there has been no follow up, and such data collection is not required by the CDC. The lack of data regarding the use of PGD reflects the lack of comprehensive regulation regarding ART. No singular agency has been tasked with oversight, and thus no one is responsible for evaluating policy effectiveness and appropriateness. Thus, this range of possibilities is just that—a list of possible uses without a clear understanding of actual practice in the United States.

It is also important to note how increasing rates of IVF and PGD reflect the larger context of the medicalization of infertility in the United States. The advancements in medical technologies that have allowed doctors to overcome physical obstacles to pregnancy have been coupled with a new intensity among patients to overcome infertility as well.[29] Maura Ryan argues that this creates a not entirely well-founded sense of optimism in the medical establishment's ability to surmount infertility, and serves to "not only generate new confidence in medicine but deepen the perception of infertility as a technical problem amenable to a technical solution."[30] In this way, pregnancy becomes treated as an illness or disability, and therefore becomes a problem that theoretically can be "cured." Clearly, advancements in reproductive technologies can be viewed as important for improving the health and well-being of both mother and child; nevertheless, they also raise ethical and moral questions.

The frame of treating pregnancy, childbirth, and genetic diseases as disorders to be "fixed" plays into the idea of the ability to reduce the risks associated with the human condition, a key theme in utopian literatures. Assisted reproductive technologies create hope for infertile individuals who desire to be parents. ART becomes the technical solution that appears to give an individual more control over reproduction. Adding to the utopian hope underlying ART cycles, Judith Daar notes that utopian desire also exists in the expectation of the result. There exists an "oft-spoken worry that ART alters parental expectations by promoting a sensibility that human intervention in reproduction produces 'healthier, better, more perfect children.'"[31] Fertility patients may perceive ART not only as the miracle solution to their infertility, but also as a method of producing "perfect" children.

Yet as Cheryl Erwin notes, "The search for stability in the form of control over reproduction and genetic endowment has the apparent advantage of avoiding uncertainty, but it is bought at the price of health risks and the possibility of not having a child with one's own egg or sperm."[32] The hope for ART is itself not inherently utopian; however, it becomes such "when we fail to reflect on the limitations of these technologies or to imagine other possible ways to work out our need to bear children and raise them."[33] In order to properly reflect on ART's limitations, patients should have access to data and a deeper understanding of potential risks and repercussions of delaying pregnancy and/or using ART. For that data to be collected, more coherent regulatory oversight is needed in the United States.

Regulatory Framework in the United States

Safety in reproductive medicine is assured by a combination of state and federal government regulation, as well as self-regulation among medical professionals.[34] The result from this multitiered approach is broad federal law coupled with a patchwork of varying state regulations more specific to procedures. Looking at ART regulation regarding a particular process, like paid surrogacy, illustrates the wide variation in state laws. For example, Louisiana state law "prohibits paid traditional surrogacy agreements, in which the surrogate provides the egg," but is "silent on gestational surrogacy agreements, in which the surrogate is not genetically related to the child, and on unpaid agreements."[35] Its neighbors' policies differ significantly: in Mississippi, "state law and case law are silent on surrogacy," while in Arkansas, "surrogacy contracts are valid and enforceable," and in Texas, "state law allows and regulates gestational surrogacy agreements," but the "intended parents must be married, and the agreement must be validated by the court."[36] Adding to the diversity of regulation at the state level are the different institutions for oversight and enforcement. In the United States, a medical board or designated agency in each state monitors physician practice and licenses individual physicians. These boards also can discipline or revoke the license of individuals who fail to uphold the law.[37]

At the federal level, three agencies have explicitly been designated to regulate ART, although the result is still piecemeal. Nationally, the CDC collects and publishes data on ART procedures; the Food and Drug Administration (FDA) controls approval and use of drugs, biological products, and medical devices, and it has jurisdiction over screening and testing of reproductive tissues, such as donor eggs and sperm; and the Centers for Medicare and Medicaid Services (CMS) is responsible for implementation of the Clinical Laboratory Improvement Act to ensure the quality of laboratory testing.[38] Thus, no one agency is responsible for policy making and oversight of ART practices in the United States. This is not unusual, per se, as the "federal government does not have direct jurisdiction over the practice of medicine" in the United States.[39] Nonetheless, without a clear locus of federal regulation, ART remains mostly in the domain of state law and self-regulation among physicians.

The lack of a comprehensive approach to regulating ART has allowed professional standards to take the place of government regulation. The American Society for Reproductive Medicine (ASRM) claims that

> the medical profession exercises significant self-regulation to assure the continuing competence of practicing physicians. Specialists in reproductive medicine are certified by the American Board of Obstetrics and Gynecology or the American Board of Urology after completing residency training and passing examinations. They may achieve subspecialty certification with additional training in infertility and endocrinology. Continuing medical education and periodic re-examination are required to maintain certification.[40]

Thus, professional standards to maintain physician certifications, overseen and implemented by other physicians on specialized boards, are supposed to ensure that doctors in obstetrics, gynecology, and urology practice safely and ethically in the United States.

From this perspective, the ASRM asserts that ART in the United States are "heavily" regulated. When considering the potential for federal law that would create oversight and regulate ART, ASRM has stated that "while properly crafted language in a widely adopted medical practice act requiring specialists in ART to follow ASRM guidelines unless otherwise indicated might improve the uniformity of practice nationwide, it is important to recognize that ART is already one of most highly regulated of all medical practices in the United States."[41] Despite the absence of a specific legislative act addressing ART, there are mandatory general medical regulations affecting reproductive medicine, including clinical ART–specific regulations, such as the Fertility Clinic Success Rate and Certification Act of 1992; nonmedical regulation of clinical ART, such as legislation related to "truth in advertising"; mandatory lab regulations and the regulation of somatic cell nuclear transfer; as well as regulation of genetic testing and treatment.

But when compared to other countries, the United States is often identified as being one of the least regulated states. In part, this is because of a lack of comprehensive legislation at the federal level that deals with ART directly, rather than the piecemeal approach to regulating separately reproductive medicine labs, clinics, processes, and advertising. As a result, the United States is often referred to as the "Wild West" of reproductive medicine, despite the ASRM's argument that the field is highly regulated.[42] Individual states may have passed laws, but the American government has not significantly stepped in. The only federal legislation passed pertaining to ART is the Fertility Clinic Success Rate and Certification Act of 1992. And regulation of ART varies at the state level, as noted above regarding the practice of paid surrogacy. Only Pennsylvania extensively regulates and monitors ART clinics and activities.[43]

A few aspects contribute to this perception that ART is largely unregulated in the United States. For one, there is no central federal agency that oversees and regulates ART. The piecemeal approach of regulation—spread across three agencies and different levels of government—allows patients to shop around for services, which in turn allows them to circumvent laws in their own states. Another issue of concern is the proliferation of clinics to meet the market demand. Some clinics are affiliated with universities or other academic centers, which means that they fall under state regulations, but most are privately owned and operated, meaning that they are largely self-regulated. Of course, clinics and doctors should be licensed under state law and report their success rates to the CDC under federal law, but while many of the guidelines set by the ASRM are reasonable, "they are not binding and are routinely flouted."[44]

Finally, there has also been a significant amount of media exposure of illegal, immoral, or irresponsible behavior, which contributes to the perception that the United States is completely unregulated.[45] And, as Rebecca Dresser argues, the perception of the United States as the "Wild West" of reproductive medicine becomes more entrenched with "each high-profile incident—advertisements offering huge sums to Ivy League students providing eggs for infertile couples, the release of new statistics on multiple births to women undergoing infertility treatment, gametes and embryos accidentally or intentionally given to the wrong patients, and the first birth following posthumous sperm retrieval."[46] These incidents, in turn, lead to calls for more regulation of the field.[47]

The case of Nadya Suleman provides an example of such a flashpoint in the public consciousness regarding reproductive technologies. Suleman gave birth to octuplets in January 2009 and within days became a media sensation.[48] The focus of the coverage, however, rapidly shifted from reporting on the rarity of an octuplet birth to probing the details of her personal life: her marital status, her financial situation, and—most tellingly—her use of ART to have a total of fourteen children.[49] Dubbed "Octomom," she was vilified in the press, a backlash that Dana-Ain Davis attributes to Suleman's failure to fit the "imagined social script" that ART is only for people in sanctioned heterosexual relationships with the economic means to raise a family.[50]

Not only does Suleman's case exemplify the implicit assumptions that Americans hold regarding ART and its use, it also illustrates an extreme case of a reproductive specialist defying professional standards. Michael Kamrava, Suleman's doctor, implanted 12 embryos into her womb, which was significantly higher (six times as much) than the suggested limit for her diagnosis.[51] Yet, Dr. Kamrava's treatment of Nadya Suleman technically was not illegal because there were no laws prohibiting this type of treatment. Nonetheless, Kamrava's license was revoked by the California Medical Board two years later, based on evidence of negligence in several patients' cases, not just Suleman's.[52]

Despite the ASRM's claims that the field is heavily regulated, the case of Nadya Suleman highlights the faults of a piecemeal regulatory framework that relies largely on self-regulation by medical professionals. As the "authors" of a utopia, specialists can help individuals suffering from infertility fulfill their desire to procreate. But the lack of comprehensive federal oversight assumes that individuals (patients and doctors alike) will choose correctly in how they use ART. Inherent in that assumption then is the notion that there is a "right" choice. Technically, ART should be available to anyone who needs it; however, the Suleman case reveals implicit bias in public perception of who should become a parent. Moreover, in approaching the body as a potentiality, the lack of regulation also assumes that the drive to cure infertility or eradicate genetic disorders is divorced from social context. Again, the Suleman case exposes how the lack of a regulatory framework allows us to avoid conversations over goals for ART and its social, cultural, and political meanings.

In the aftermath of the "Octomom" story, California State Senator Gloria Negrete McLeod introduced legislation accreditation standards and guidelines for the operation of fertility clinics and surgical centers, but it was vetoed by Governor Schwarzenegger.[53] Subsequent efforts to create more oversight also failed in the California legislature. Although the failure to develop comprehensive legislation could appear to be a purposeful position for politicians—why extend access to fertility treatments to parts of the population that are struggling economically?—in general, the absence of oversight results from the contentious state of reproductive issues in the United States. Political arguments over the status of the embryo, stem cell research, and abortion have hampered public policy making for ART.[54] Links made between PGD and abortion especially inhibit the ability of policy makers to discuss potential regulations, with both procedures evoking concerns of moral ambiguity and conflict.[55] In particular, ethical issues arise over the process of PGD because it involves the creation of several embryos with the understanding that not all will be implanted and some will be destroyed.[56] Thus, any potential regulation of ART might rely on abortion regulation as a base,[57] which could open the door further to religious opposition to ART and possibly shut down innovation in the field.[58]

Ethical and Regulatory Challenges

Although the context of American controversy over reproductive issues and the status of the embryo may contribute to why the field is left to self-regulation through professional associations, the use of biotechnology in ART—and specifically PGD—raises other ethical and regulatory challenges. In the immediate term, several ethical dilemmas are apparent. For example, is it ethical to siphon off and test the cells that can develop into human life? Do potential parents have the right to manipulate their genetic material? Do they have the right to implant embryos knowing they may then selectively abort? Should potential parents be able to choose embryos to implant based on sex? But there are also long-term, potentially dystopian outcomes of these practices in both the social and economic spheres.

Some of the feared social implications of the expanding use of PGD include: shifting expectations for parent and child; increased ostracism of people with disabilities; and reproductive stratification. As mentioned above, the determined optimism with which infertile patients may view ART could result in the perception that these procedures can produce "better" children, leading to shifts in parental expectations regarding the health and abilities of a child. In addition, knowledge of the means of conception might destabilize an individual's self-worth and identity by engendering questions about his or her origins.[59] The use of ART more broadly impacts expectations regarding the role of women in society and motherhood, as well. The presentation of ART as a technical solution to infertility could reinforce the idea that a woman must have children in order to be socially valuable or fulfilled, no matter her age or health status. In this way, ART reduces women to their reproductive capacities,[60] an issue of particular concern for women's rights advocates.

Disability advocates have also expressed apprehension about the use of PGD. The ability of PGD to detect conditions of varying degrees, from debilitating to non–life threatening, opens the door to embryo selection based on aesthetic concerns, convenience, or prejudice.[61] Procedures to prevent the birth of disabled persons then become "inseparably tied to eugenics and can only reinforce the worst stereotypes of people with disabilities,"[62] especially if "an attitude develops that 'certain diseases can and therefore should be avoided.'"[63] Furthermore, fears about developing techniques that could employ genetic transfers with IVF and PGD raise the specter of a "post-human future" where people will mix and match genes.[64] As Heather Long notes in an editorial on PGD, "When the world looks back at how the 'designer babies' trend began, they will see an innocent start . . . [Genetic screening] will likely be a progression from just wanting a child, to wanting one less likely to get certain diseases, to wanting one more likely to have traits associated with being taller or more artistic or athletic."[65] Even without the possibility of genetic transfer techniques and designer babies, widespread use of PGD for sex selection could result in a gender imbalance, potentially reinforcing the cultural privileging of male children, and devaluing women.[66] Thus, despite PGD's auspicious beginnings as a remedy for genetic anomalies, its potential uses for more morally ambiguous goals demand a closer regulatory look.

The social impacts of PGD are inherently entwined with the economic issues surrounding its usage, especially in the context of reproductive stratification. The expense of these technologies, and the lack of insurance support, currently means that certain portions of the population have consistent access to these procedures while others are excluded. Only 15 states in the United States have passed legislation requiring some form of insurance coverage for fertility treatments, although the type of coverage widely varies.[67] Without assured coverage, potential parents face financial barriers to equal access. The average IVF cost in the United States is $12,400,[68] but it can be as much as $15,000.[69] These prices are for one cycle of IVF, and do not include additional expenses such as PGD (about $3,550),[70] the cost of a frozen embryo transfer, embryo freezing and storage, using egg or sperm donors, or other associated treatments. There is no regulatory limit to the number of IVF cycles a woman may choose to undergo, although three cycles is the recommended amount to improve the woman's chances of getting pregnant.[71] As a result of these costs, groups that have the financial means (upper-/middle-class whites with access to healthcare) are more likely to pursue treatment to deal with infertility.[72]

The long-term consequences of this inequity to fertility treatment access reinforce perceptions of not only who is fit to be a parent, but also who should be a mother.[73] Moreover, continually expanding use of the procedure among only those who have the financial means raises the possibility of class stratification based on genetics. In this dystopian future, elites will be able to produce "perfected" children, free of genetic disorders and with physical and/or cognitive "enhancements." These genetic benefits will strengthen the advantages already enjoyed by elites, creating more deeply entrenched social and economic divisions.

Is Self-Regulation Sufficient?

Although genetic engineering procedures for physical and cognitive enhancements are still just a prediction for the future, the currently available procedure PGD is without legal restriction in the United States. ASRM argues that the field is already highly regulated and professional standards are sound; yet, its claims fall short when looking at recent data collected by the CDC. As Kirsten Riggan notes, the majority of reproductive health clinics in the United States are members of professional organizations such as the Society for Assisted Reproductive Technology (SART) or ASRM, but not all follow the clinical and ethical guidelines conveyed by these organizations.[74] For example, in 2009 the Practice Committees of both organizations jointly issued a statement regarding the criteria to determine the limit of embryos that should be transferred.[75] The guidelines outlined the criteria for a "favorable" prognosis[76] and the limit of the number of embryos that should be implanted by age group in favorable or less favorable situations. In general, ASRM and SART recommended one to two embryos for transfer in women who are 35 years or younger and have a favorable prognosis.[77]

But the CDC's most recent data indicate that some women with favorable diagnoses were implanted with three, four, or more embryos in one IVF cycle.[78] The guidelines emphasize the importance of singleton live births because they have a much lower risk than multiple-infant births of complications with maternal health and of adverse infant health outcomes, such as prematurity, low birth weight, disability, and death.[79] This emphasis has carried over into ASRM's and SART's more recent guidelines, which decrease the suggested limit in women aged 35 or younger to one embryo for implantation, in an effort to reduce the number of multiple births that have occurred because of lax adherence to the 2009 guidelines.[80]

A 2006 survey of ART clinics by Johns Hopkins also reveals the flouting of the professional standards with regard to the use of PGD. ASRM's Ethics Committee published standards in 1999 on the use of PGD for sex selection. The Ethics Committee acknowledged that "to encourage PGD for sex selection when it is not medically indicated presents ethical problems," and that the widespread use of PGD for sex selection could result in an imbalance in society's gender ratio and negative consequences for obtaining gender equality.[81] Despite these standards, the results of the Johns Hopkins survey revealed that 9 percent of all PGD cycles were for nonmedical sex selection purposes.[82] Moreover, of the 137 clinics identified as providing PGD-IVF services, 42 percent offered PGD as a service for nonmedical sex selection.[83] ESHRE PGD Consortium's data also show that "social sexing," as they label it, was the motivation of 671 PGD cycles reported between 1997 and 2007, with the vast majority of those originating from one clinic in the United States.[84] Clearly, self-regulation only works when the rules and standards are consistently and extensively followed.

Conclusion

Even though the data seem to indicate practitioner indifference to the guidelines laid out by ASRM and SART, it is difficult to say how widespread the issue may be because the scale is small: only about 5 percent of all ART cycles performed use PGD. However, for the most recent data, that still translates to over 8,000 cycles in 2012 in which the method was used. Unfortunately, the CDC does not indicate how many of these cycles resulted in live births. We might assume a higher success rate than the overall clinical success rate because the use of PGD should eliminate embryos that suffer from obvious abnormalities that would prevent them from developing in the womb and result in failed

IVF cycles. Nevertheless, without a mandate to collect the data, clinics do not report PGD success rates.

At such a small number, fears of the dystopian future outlined above seem possible only far into the future. But now is the time to address these concerns, before the practice becomes more widespread and/or more exclusive. The lack of government regulation implicitly allows for these types of practices to proliferate. So, although the American government is not exercising control over citizens' bodies by making laws directly determining which types of people are "worthy" of ART, its failure to act in a comprehensive and a cohesive manner to deal with these regulatory challenges allows the inequalities to persist. Absence from the market shapes the standards that exist, and thus indirectly allows the "right" kinds of people to reproduce. Americans with money, education, and access to healthcare will determine the trends in reproductive medicine and procedures. The current regulatory system is based on the utopian hope that people will use genetic testing for good, however that may be defined. But we know already that reproductive specialists are not following the ethical guidelines laid out by their own professional associations and experts. We cannot believe that potential parents will be any better at adhering to ethical considerations, especially not when the welfare of their children is at stake.

Notes

1. Ruth Levitas, *Ralahine Utopian Studies, Volume 3: Concept of Utopia* (Brussels, Belgium: Peter Lang AG, 2010), 2.
2. Levitas, *Ralahine Utopian Studies*, 6.
3. Ibid.
4. Bryan S. Turner, *Regulating Bodies: Essays in Medical Sociology* (New York: Routledge, 2002), 16.
5. Turner, *Regulating Bodies*, 16.
6. Levitas, *Ralahine Utopian Studies*, 1.
7. CDC, "What Is Assisted Reproductive Technology?" Accessed May 11, 2014, http://www.cdc.gov/art/. Emphasis in original.
8. CDC, "What is Assisted Reproductive Technology?"
9. Data for this table was taken from the CDC's website: CDC, "2010 ART National Summary Report," accessed May 11, 2014, http://www.cdc.gov/art/ART2010/PDFs/ART_2010_National_Summary_Report.pdf; CDC, "2011 ART National Summary Report," accessed May 11, 2014, http://www.cdc.gov/art/ART2011/PDFs/ART_2011_National_Summary_Report.pdf; and CDC, "What Is Assisted Reproductive Technology?"
10. CDC, "What Is Assisted Reproductive Technology?"

11. CDC, "PGD (Preimplantation Genetic Diagnosis)," accessed May 13, 2014, http://www.cdc.gov/art/ART2010/appixb.htm#P.
12. Gwendolyn P. Quinn et al., "Frozen Hope: Fertility Preservation for Women with Cancer," *Journal of Midwifery & Women's Health*, 55.2 (2010): 175–180, accessed May 13, 2014, http://www.medscape.com/viewarticle/717970_3.
13. Quinn et al., "Frozen Hope."
14. Judith F. Daar, "ART and the Search for Perfectionism: On Selecting Gender, Genes, and Gametes," *The Journal of Gender, Race, and Justice*, 9 (2005): 248.
15. Marcia C. Inhorn, and Daphne Birenbaum-Carmeli, "Assisted Reproductive Technologies and Culture Change," *Annual Review of Anthropology*, 37 (2008): 185.
16. Daar, "ART and the Search for Perfectionism," 248.
17. Ibid., 249.
18. The President's Council on Bioethics, "Reproduction and Responsibility: The Regulation of New Biotechnologies," accessed May 13, 2014, https://bioethicsarchive.georgetown.edu/pcbe/reports/reproductionandresponsibility/chapter3.html.
19. Susannah Baruch, David Kaufman, and Kathy L. Hudson, "Genetic Testing of Embryos: Practices and Perspectives of US In Vitro Fertilization Clinics," *Fertility and Sterility* 89.5 (2008), accessed May 13, 2014, http://www.dnapolicy.org/resources/GeneticTestingofEmbryos.pdf.
20. Reproductive Health Technologies Project, "Pre-implantation Genetic Diagnosis (PGD)," accessed May 13, 2014, http://www.rhtp.org/fertility/pgd/.
21. CDC, "National ART Success Rates," accessed May 13, 2014, http://nccd.cdc.gov/DRH_ART/Apps/NationalSummaryReport.aspx.
22. Data for this analysis was taken from: CDC, "What Is Assisted Reproductive Technology?"; and CDC, "Preliminary 2012 Data—Clinic Tables and Data Dictionary," accessed May 19, 2014, http://www.cdc.gov/art/artreports.htm.
23. Data was collected from the "Preliminary 2012 Data—Clinic Tables and Data Dictionary"; percentages and formulas are the author's own.
24. While the 2012 preliminary data includes rates of PGD usage from individual reporting clinics, the CDC has not yet provided a preliminary or definitive average rate of PGD usage across all reporting clinics.
25. Assuming the success rate as consistent with the overall average might be conservative, as the goal of PGD is partially to improve the chances of implanting a viable embryo by identifying embryos without the defects that might lead to non-implantation in the womb or spontaneous abortion. However, a study published in 2006 showed that "PGD has not been shown to increase pregnancy rates" in high-risk populations, such as women of advanced maternal age (Lora K. Shahine, and Marcelle I. Cedars, "Preimplantation Genetic Diagnosis Does Not Increase Pregnancy Rates in Patients at Risk for Aneuploidy," *Fertility and Sterility* 85.1 (2006): 54).
26. J. C. Harper et al., "The ESHRE PGD Consortium: 10 Years of Data Collection," *Human Reproduction Update* (2012), accessed June 16, 2014, http://www.

eshre.eu/~/media/emagic%20files/Data%20collection/PGD/PGD%20Consortium/10%20yrs%20manuscript.pdf.
27. Harper et al., "The ESHRE PGD Consortium," 10. A total of 87 clinics reported data to the ESHRE between 1997 and 2007, although not necessarily every year. From across Europe, 56 clinics submitted data; from North and South America, 11; and from Africa, Asia, and Australia, 20.
28. Baruch et al. "Genetic Testing of Embryos."
29. Maura A. Ryan, *Ethics and Economics of Assisted Reproduction* (Washington, DC: Georgetown University Press, 2001), 76.
30. Ryan, *Ethics and Economics of Assisted Reproduction*, 76.
31. Daar, "ART and the Search for Perfectionism," 244.
32. Cheryl Erwin, "Utopian Dreams and Harsh Realities: Who Is in Control of Assisted Reproductive Technologies in a High-Tech World?" *The Journal of Gender, Race, and Justice* 9 (2006): 625.
33. Erwin, "Utopian Dreams and Harsh Realities," 623.
34. ASRM, "Oversight of Assisted Reproductive Technology," accessed May 15, 2014, http://www.asrm.org/uploadedFiles/Content/About_Us/Media_and_Public_Affairs/OversiteOfART%20(2).pdf.
35. Alison Sherwood, "Surrogacy Laws in the United States," *Milwaukee Journal Sentinel*, August 4, 2012, accessed May 31, 2014, http://www.jsonline.com/news/health/163772546.html.
36. Sherwood, "Surrogacy laws in the United States."
37. ASRM, "Oversight of Assisted Reproductive Technology," 3.
38. Ibid.
39. Genetics and Public Policy Center, "Reproductive Genetic Testing: A Regulatory Patchwork [United States]," accessed May 31, 2014, http://www.dnapolicy.org/policy.international.php?action=detail&laws_id=63.
40. ASRM, "Oversight of Assisted Reproductive Technology," 3.
41. Ibid.
42. For example, see: Meredith Leigh Birdsall, "An Exploration of 'The 'Wild West' of Reproductive Technology': Ethical and Feminist Perspectives on Sex-Selection Practices in the United States," *William & Mary Journal of Women and the Law* 17.1 (2010): 223–247; Francine Coeytaux, Marcy Darnovsky, and Susan Berke Fogel, "Assisted Reproduction and Choice in the Biotech Age: Recommendations For a Way Forward," *Contraception Journal* (Jan 2011), accessed June 4, 2014, http://www.arhp.org/publications-and-resources/contraception-journal/january-2011; Alexander N. Hecht, "The Wild Wild West: Inadequate Regulation of Assisted Reproductive Technology," *Houston Journal of Health Law and Policy*, 1 (2001): 227–261; or Kimberly M. Mutcherson, "Welcome to the Wild West: Protecting Access to Cross Border Fertility Care in the United States," *Cornell Journal of Law and Public Policy*, 22.2 (2012): 349–393.
43. Kirsten Riggan, "G12 Country Regulations of Assisted Reproductive Technologies," The Center for Bioethics and Human Dignity, October 1, 2010, accessed May 31, 2014, http://cbhd.org/content/g12-country-regulations-assisted-reproductive-technologies.

44. Coeytaux, Darnovsky, and Berke Fogel. The authors note, for example, that surveys reveal that the majority of clinics do not follow the ASRM's guidelines to transfer only one embryo at a time when treating women under 35 with a favorable prognosis. In addition, over 40 percent of polled clinics offered nonmedical sex-selection services through PGD, although it is a practice discouraged by the ASRM (Baruch et al., "Genetic Testing of Embryos.").
45. David Adamson, "Regulation of Assisted Reproductive Technologies in the United States," *Family Law Quarterly* 39.3 (2005): 728.
46. Rebecca Dresser, "Regulating Assisted Reproduction," *The Hastings Center Report*, 30.6 (2000): 26.
47. Dresser, "Regulating Assisted Reproduction," 26.
48. Dana-Ain Davis, "The Politics of Reproduction: The Troubling Case of Nadya Suleman and Assisted Reproductive Technology," *Transforming Anthropology* 17.2 (2009): 105–116.
49. Davis, "The Politics of Reproduction," 107–108.
50. Ibid., 108.
51. Associated Press, "Octomom's Fertility Doctor Wants to Treat Patients," *The Huffington Post*, November 23, 2011, accessed June 11, 2014, http://www.huffingtonpost.com/2011/11/23/octomoms-fertility-doctor_n_1109579.html. Initial reports indicated that six embryos had been transferred and two had split into twins (see, for example: Alison Stateman, "The Fertility Doctor behind the 'Octomom,'" *TIME Magazine*, March 7, 2009, accessed June 11, 2014, http://content.time.com/time/nation/article/0,8599,1883663,00.html), but a review of Dr. Kamrava's records during a licensing hearing revealed that he had, in fact, implanted 12 embryos (AP).
52. Associated Press, "Octomom's fertility doctor has license revoked," *CBS News*, June 2, 2011, accessed June 11, 2014, http://www.cbsnews.com/news/octomoms-fertility-doctor-has-license-revoked/.
53. "Close loopholes in clinic oversight," *The Daily Times*, April 24, 2010, accessed June 19, 2014, http://www.daily-times.com/farmington-business/cI_14953782.
54. Erik Parens, and Lori P. Knowles, "Special Supplement: Reprogenetics and Public Policy: Reflections and Recommendations," *The Hastings Center Report*, 33.4 (2003): S10.
55. Kimberly M. Mutcherson, "Making Mommies: Law, Pre-implantation Genetic Diagnosis, and the Complications of Pre-motherhood," *Columbia Journal of Gender and Law*, 18.1 (2008): 352.
56. Mutcherson, "Making Mommies," 352.
57. Beth A. Burkstrand-Reid, "The More Things Change…: Abortion Politics and the Regulation of Assisted Reproductive Technology," *UMKC Law Review*, Winter (2010): 367.
58. June Carbone, "Negating the Genetic Tie: Does the Law Encourage Unnecessary Risks?" *UMKC Law Review*, Winter (2010): 354.
59. Timothy F. Murphy, "Access and Equity: International Standards and Assisted Reproductive Technologies," *Ethics, Law and Moral Philosophy of Reproductive Biomedicine*, 2.1 (2007): 14.

60. Murphy, "Access and Equity," 14; Ryan, *Ethics and Economics of Assisted Reproduction*, 53.
61. Kathy L. Hudson, "Preimplantation Genetic Diagnosis: Public Policy and Public Attitudes," *Fertility and Sterility* 85.6 (2006): 1642.
62. Murphy, "Access and Equity," 14.
63. Daar, "ART and the Search for Perfectionism," 260.
64. Murphy, "Access and Equity," 14; Francis Fukuyama, *Our Posthuman Future: Consequences of the Biotechnology Revolution* (New York: Farrar, Straus, and Giroux, 2002).
65. Heather Long, "'Designer Babies': The Ultimate Privileged Elite?" *The Guardian*, July 9, 2013, accessed June 19, 2014, http://www.theguardian.com/commentisfree/2013/jul/09/ivf-baby-born-genetic-selection-ultimate-elite.
66. Hudson, "Preimplantation Genetic Diagnosis," 1643.
67. ASRM, "State Infertility Insurance Laws," accessed June 11, 2014, http://www.asrm.org/insurance.aspx.
68. ASRM, "Q06: Is In Vitro Fertilization Expensive?" accessed June 11, 2014, http://www.asrm.org/detail.aspx?id=3023.
69. Resolve: The National Infertility Association, "The Cost of Infertility Treatment," accessed June 11, 2014, http://www.resolve.org/family-building-options/insurance_coverage/the-costs-of-infertility-treatment.html.
70. Resolve: The National Infertility Association, "The Cost of Infertility Treatment."
71. Resolve, "Why IVF Success Rates Matter to You," accessed June 11, 2014, http://www.resolve.org/family-building-options/why-ivf-success-rates-matter-to-you.html.
72. Davis, "The Politics of Reproduction," 110.
73. Davis uses the backlash against Nadya Suleman ("Octomom") as an illustration of the intrinsic assumptions of the American public of ART being a privilege for white heterosexual couples (111).
74. Kirsten Riggan, "Regulation (or Lack Thereof) of Assisted Reproductive Technologies in the U.S. and Abroad," The Center for Bioethics and Human Dignity, March 5, 2011, accessed June 11, 2014, https://www.donorsiblingregistry.com/sites/default/files/files/Regulation%20(or%20Lack%20Thereof)%20of%20ART.pdf.
75. "Guidelines on the Number of Embryos Transferred," *Fertility and Sterility* 92.5 (2009): 1518–1519.
76. A "favorable" diagnosis may result from the following conditions: the IVF cycle is the woman's first; the embryos are good quality; there are excess embryos available for cryopreservation; or the woman has previously undergone a successful IVF cycle; diagnoses that do not fit any of these categories are labeled "all others" ("Guidelines on the Number of Embryos Transferred").
77. "Guidelines on the Number of Embryos Transferred," 1519.
78. CDC, "2011 ART National Summary Report," 32.
79. "Guidelines on the Number of Embryos Transferred," 1518; CDC, "2011 ART National Summary Report," 12.

80. "Criteria for Number of Embryos to Transfer A Committee Opinion," *Fertility and Sterility* 99 (2013): 44–46.
81. "Sex Selection and Preimplantation Genetic Diagnosis," *Fertility and Sterility* 72.4 (1999): 595–598, accessed May 11, 2014, http://www.asrm.org/uploadedFiles/ASRM_Content/News_and_Publications/Ethics_Committee_Reports_and_Statements/Sex_Selection.pdf.
82. Baruch et al., "Genetic Testing of Embryos," 3–5.
83. Ibid.
84. Harper et al., "The ESHRE PGD Consortium," 10.

Bibliography

Adamson, David. "Regulation of Assisted Reproductive Technologies in the United States." *Family Law Quarterly* 39.3 (2005): 727–744.

American Society for Reproductive Medicine (ASRM). "Oversight of Assisted Reproductive Technology." Accessed May 15, 2014. http://www.asrm.org/uploadedFiles/Content/About_Us/Media_and_Public_Affairs/Oversite OfART%20(2).pdf.

———. "Q06: Is In Vitro Fertilization Expensive?" Accessed June 11, 2014. http://www.asrm.org/detail.aspx?id=3023.

———. "State Infertility Insurance Laws." Accessed June 11, 2014. http://www.asrm.org/insurance.aspx.

Associated Press. "Octomom's Fertility Doctor Has License Revoked." *CBS News*, June 2, 2011. Accessed June 11, 2014. http://www.cbsnews.com/news/octomoms-fertility-doctor-has- license-revoked/.

———. "Octomom's Fertility Doctor Wants to Treat Patients." *The Huffington Post*, November 23, 2011. Accessed June 11, 2014. http://www.huffingtonpost.com/2011/11/23/octomoms-fertility-doctor_n_1109579.html.

Baruch, Susannah, David Kaufman, and Kathy L. Hudson. "Genetic Testing of Embryos: Practices and Perspectives of US In Vitro Fertilization Clinics." *Fertility and Sterility* 89.5 (2008): 1053–1058. Accessed May 13, 2014. http://www.dnapolicy.org/resources/GeneticTestingofEmbryos.pdf.

Birdsall, Meredith Leigh. "An Exploration of 'The 'Wild West' of Reproductive Technology': Ethical and Feminist Perspectives on Sex-Selection Practices in the United States." *William & Mary Journal of Women and the Law* 17.1 (2010): 223–247.

Burkstrand-Reid, Beth A. "The More Things Change…: Abortion Politics and the Regulation of Assisted Reproductive Technology." *UMKC Law Review* 79 (Winter, 2010): 361–372.

Carbone, June. "Negating the Genetic Tie: Does the Law Encourage Unnecessary Risks?" *UMKC Law Review* 79 (Winter, 2010): 333–360.

Centers for Disease Control and Prevention (CDC). "2010 ART National Summary Report." Accessed May 11, 2014. http://www.cdc.gov/art/ART2010/PDFs/ART_2010_National_Summary_Report.pdf.

———. "2011 ART National Summary Report." Accessed May 11, 2014. http://www.cdc.gov/art/ART2011/PDFs/ART_2011_National_Summary_Report.pdf.

———. "National ART Success Rates." Accessed May 13, 2014. http://nccd.cdc.gov/DRH_ART/Apps/NationalSummaryReport.aspx.

———. "PGD (Preimplantation Genetic Diagnosis)." Accessed May 13, 2014. http://www.cdc.gov/art/ART2010/appixb.htm#P.

———. "Preliminary 2012 Data—Clinic Tables and Data Dictionary." Accessed May 19, 2014. http://www.cdc.gov/art/artreports.htm.

———. "What is Assisted Reproductive Technology?" Accessed May 11, 2014. http://www.cdc.gov/art/.

Coeytaux, Francine, Marcy Darnovsky, and Susan Berke Fogel. "Assisted Reproduction and Choice in the Biotech Age: Recommendations for a Way Forward." *Contraception Journal* (2011). Accessed June 4, 2014. http://www.arhp.org/publications-and-resources/contraception-journal/january-2011.

"Close Loopholes in Clinic Oversight." *The Daily Times*, April 24, 2010. Accessed June 19, 2014. http://www.daily-times.com/farmington-business/cI_14953782.

"Criteria for Number of Embryos to Transfer: A Committee Opinion." *Fertility and Sterility* 99 (2013): 44–46.

Daar, Judith F. "ART and the Search for Perfectionism: On Selecting Gender, Genes, and Gametes." *The Journal of Gender, Race, and Justice* 9 (2005): 242–272.

Davis, Dana-Ain. "The Politics of Reproduction: The Troubling Case of Nadya Suleman and Assisted Reproductive Technology." *Transforming Anthropology* 17.2 (2009): 105–116.

Dresser, Rebecca. "Regulating Assisted Reproduction." *The Hastings Center Report* 30.6 (2000): 26–27.

Erwin, Cheryl. "Utopian Dreams and Harsh Realities: Who Is in Control of Assisted Reproductive Technologies in a High-Tech World?" *The Journal of Gender, Race, and Justice* 9 (2006): 621–635.

Fukuyama, Francis. *Our Posthuman Future: Consequences of the Biotechnology Revolution*. New York: Farrar, Straus, and Giroux, 2002.

Genetics and Public Policy Center. "Reproductive Genetic Testing: A Regulatory Patchwork [United States]."Accessed May 31, 2014. http://www.dnapolicy.org/policy.international.php?action=detail&laws_id=63.

"Guidelines on the Number of Embryos Transferred." *Fertility and Sterility* 92.5 (2009): 1518–1519.

Harper, J. C., L. Wilton, J. Traeger-Synodinos, V. Goossens, C. Moutou, S. B. SenGupta, T. Pehlivan Budak, P. Renwick, M. De Rycke, J. P. M. Geraedts, and G. Harton. "The ESHRE PGD Consortium: 10 years of data collection." *Human Reproduction Update* (2012): 1–12. Accessed June 16, 2014. http://www.eshre.eu/-/media/emagic%20files/Data%20collection/PGD/PGD%20Consortium/10%20yrs%20manuscript.pdf.

Hecht, Alexander N. "The Wild Wild West: Inadequate Regulation of Assisted Reproductive Technology." *Houston Journal of Health Law and Policy* 1 (2001): 227–261.

Hudson, Kathy L. "Preimplantation Genetic Diagnosis: Public Policy and Public Attitudes." *Fertility and Sterility* 85.6 (2006): 1638–1645.

Inhorn, Marcia C., and Daphne Birenbaum-Carmeli. "Assisted Reproductive Technologies and Culture Change." *Annual Review of Anthropology* 37 (2008): 177–196.

Levitas, Ruth. *Ralahine Utopian Studies, Volume 3: Concept of Utopia.* Brussels, Belgium: Peter Lang AG, 2010.

Long, Heather. "'Designer Babies': The Ultimate Privileged Elite?" *The Guardian*, July 9, 2013. Accessed June 19, 2014. http://www.theguardian.com/commentisfree/2013/jul/09/ivf-baby-born-genetic-selection-ultimate-elite.

Murphy, Timothy F. "Access and Equity: International Standards and Assisted Reproductive Technologies." *Ethics, Law and Moral Philosophy of Reproductive Biomedicine* 2.1 (2007): 12–18.

Mutcherson, Kimberly M. "Making Mommies: Law, Pre-implantation Genetic Diagnosis, and the Complications of Pre-motherhood." *Columbia Journal of Gender and Law* 18.1 (2008): 313–390.

———. "Welcome to the Wild West: Protecting Access to Cross Border Fertility Care in the United States." *Cornell Journal of Law and Public Policy* 22.2 (2012): 349–393.

Parens, Erik, and Lori P. Knowles, "Special Supplement: Reprogenetics and Public Policy: Reflections and Recommendations," *The Hastings Center Report* 33.4 (2003): S1–S24.

The President's Council on Bioethics. "Reproduction and Responsibility: The Regulation of New Biotechnologies." Accessed May 13, 2014. https://bioethicsarchive.georgetown.edu/pcbe/reports/reproductionandresponsibility/chapt er3.html.

Quinn, Gwendolyn P., Susan T. Vadaparampil, Paul B. Jacobsen, Caprice Knapp, David L. Keefe, Geri E. Bell, and Moffitt Fertility Preservation Group. "Frozen Hope: Fertility Preservation for Women with Cancer." *Journal of Midwifery & Women's Health* 55.2 (2010): 175–180. Accessed May 13, 2014. http://www.medscape.com/viewarticle/717970_3.

Reproductive Health Technologies Project. "Pre-implantation Genetic Diagnosis (PGD)." Accessed May 13, 2014. http://www.rhtp.org/fertility/pgd/.

Resolve: The National Infertility Association. "The Cost of Infertility Treatment." Accessed June 11, 2014. http://www.resolve.org/family-building-options/insurance_coverage/the-costs-of-infertility-treatment.html.

———. "Why IVF Success Rates Matter to You." Accessed June 11, 2014. http://www.resolve.org/family-building-options/why-ivf-success-rates-matter-to-you.html.

Riggan, Kirsten. "G12 Country Regulations of Assisted Reproductive Technologies." The Center for Bioethics and Human Dignity, October 1, 2010. Accessed May 31, 2014. http://cbhd.org/content/g12-country-regulations-assisted-reproductive-technologies.

———. "Regulation (or Lack Thereof) of Assisted Reproductive Technologies in the U.S. and Abroad." The Center for Bioethics and Human Dignity, March 5,

2011. Accessed June 11, 2014. https://www.donorsiblingregistry.com/sites/default/files/files/Regulation%20(or%20Lack%20Thereof)%20of%20ART.pdf.

Ryan, Maura A. *Ethics and Economics of Assisted Reproduction*. Washington, DC: Georgetown University Press, 2001.

"Sex Selection and Preimplantation Genetic Diagnosis." *Fertility and Sterility* 72.4 (1999): 595–598. Accessed May 11, 2014. http://www.asrm.org/uploadedFiles/ASRM_Content/News_and_Publications/Ethics_Co mmittee_Reports_and_Statements/Sex_Selection.pdf.

Shahine, Lora K., and Marcelle I. Cedars. "Preimplantation Genetic Diagnosis Does Not Increase Pregnancy Rates in Patients at Risk for Aneuploidy." *Fertility and Sterility* 85.1 (2006): 51–56.

Sherwood, Alison. "Surrogacy Laws in the United States," *Milwaukee Journal Sentinel*, August 4, 2012. Accessed May 31, 2014. http://www.jsonline.com/news/health/163772546.html.

Stateman, Alison. "The Fertility Doctor behind the 'Octomom.'" *TIME Magazine*, March 7, 2009. Accessed June 11, 2014. http://content.time.com/time/nation/article/0,8599,1883663,00.html.

Turner, Bryan S. *Regulating Bodies: Essays in Medical Sociology*. New York: Routledge, 2002.

PART II

Speculations

CHAPTER 4

Utopian Visions of "Making People": Science Fiction and Debates on Cloning, Ectogenesis, Genetic Engineering, and Genetic Discrimination

Evie Kendal

Introduction: Definitions of Utopia

In his treatise on science fiction (sf), Darko Suvin claims utopian fiction as the "socio-political subgenre of science fiction," suggesting that such texts should be approached both as literature and as social scientific experiments.[1] Applying to them the label "social-science-fiction," Suvin treats the "what if" thought experimentation involved in the creation of fictional utopian worlds as being innately political.[2] For Suvin, the study of utopian fiction must begin with an exploration of collective psychology and end with a discussion of the "politics of the human species," exploring such politico-eschatological questions as to how the human species can "survive and humanize its segment of the universe."[3] While this is a useful starting point, this essay will adopt the revised definition of Suvin's "utopian fiction" put forward by such scholars as Lyman Tower Sargent, which encompasses both fictional eutopias (ideal worlds) and dystopias (bad worlds).[4] For Sargent, utopian sf is a form of "social dreaming" in which authors create alternate worlds in order to explore the social and political lives of their inhabitants.[5] Eutopias represent pleasant dreams—dystopias, nightmares.

In their "Utopias" entry for *The Encyclopedia of Science Fiction*, Brian M. Stableford and David Langford claim that all utopias deal with "hypothetical sociology and political science," thereby possessing an inherently speculative nature that justifies their classification as works of sf.[6] Claiming sf as the "West's mythform," Kieran Tranter further defines it as "the dreaming site for the West's technological futures, a place for working through both the anxieties and promises of technological change."[7] As the sf genre serves as an exploration of the scientific imaginary, the sociopolitical motivations behind its use in debates regarding emerging science and technologies warrants critical examination. In this chapter, I argue that while there is great potential for utopian sf to promote balanced discussion of the ramifications of technological development on society, at present its use predominantly favors a socially conservative political agenda. I achieve this through first addressing where utopian fiction meets bioethics and biopolitics, before moving on to discuss how various stakeholders use utopian sf in debates surrounding emerging technologies. Some examples of leading sf texts that have negatively portrayed certain technologies are considered along with the impact such representations have had on the development and integration of these technologies in society.

The Intersection between Utopian Fiction, Bioethics, and Biopolitics

The study of bioethics is one area that relies heavily on the language of utopian sf literature and film to debate such issues as the meaning of life, the nature of humanity, and the ethical concerns surrounding certain medical and technological advances. As such, it is important to consider what sf achieves for the rapidly expanding field of bioethics, both in terms of the potential for providing accessible philosophical arguments for public debate, and the risks of fueling sensationalist or negatively prejudiced images of emerging technologies. In this chapter, I explore the impact of sf imagery on the bioethical debates surrounding cloning, ectogenesis, genetic engineering, and genetic discrimination, paying particular attention to references to Aldous Huxley's *Brave New World* (1932), Mary Shelley's *Frankenstein* (1818) and Andrew Niccol's *Gattaca* (1997). Existing both as philosophical thought experiments and as socially conservative, technophobic cautionary tales, these texts serve a distinct political agenda when invoked in bioethics literature. Sherryl Vint claims that biopolitics and sf speculation converge at the point at which political governance extends into biological life, with the intent of steering the future of that life.[8] Huxley's, Shelley's, and Niccol's texts all represent leading examples of such a convergence in sf, relying on sf genre tropes to dramatize the potential threats science and technology may pose to humanity's freedom. The hubristic scientist "playing God," science run amok, the inadvertent creation of monsters, and life under biopsychosocial totalitarianism are all effective propaganda against certain areas of scientific discovery, particularly reproductive biotechnology.

The three texts above were chosen because they are among the most frequently referenced in bioethics scholarship. They also explore fictional worlds in which reproductive biotechnologies have fundamentally altered perceptions of human biology and reproduction. In *Brave New World*, widespread use of cloning and ectogenesis has almost completely replaced biological gestation (the exceptions being the naturally born "savages" in the reservations). Likewise, in *Gattaca*, every member of the fictional society must either engage in genetic engineering or have their offspring suffer the consequences of their failure to do so, including genetic discrimination in the workplace. Often heralded as the first sf novel, Shelley's classic horror *Frankenstein* functions in this essay as an example of a proto-dystopia, a small scale disruption to the world as we know it, but one possessing the potential to have global consequences.[9] Although the immediate horror of Frankenstein's monster is contained within the doctor's personal interactions with him in the novel, the real cause for alarm is the potential for the monster to change the "natural" balance of power in the world beyond this narrow sphere. In the end, Dr. Victor Frankenstein does not actually create a "race of devils"; however, the possibility of such a future race arising and challenging the supremacy of the human species remains a constant preoccupation throughout the novel.[10] When it appears in bioethical debates, *Frankenstein* is often invoked in a similar manner to *Brave New World* and *Gattaca*, and always to advance a particular sociopolitical agenda. More often than not, such agendas are distinctly anti-technological.[11]

With regard to their biopolitical significance, these three texts are not only concerned with who should have power over other people in general, but also more specifically who should have the authority to direct the future of human biology. In *Frankenstein*, an individual scientist has usurped both God and mother nature; in *Brave New World*, the whole of human biology has fallen under the strict regulation of the state; and in *Gattaca*, the rule of the majority has institutionalized genetic discrimination through the normalization of population-wide genetic engineering programs.[12] Each text also demonstrates the struggle for power over life and death that Michel Foucault identifies in *The History of Sexuality*. This power is imposed through regulatory control over the "species body" and its reproductive functions, or what Foucault terms the *biopolitics of the population*.[13] Dr. Frankenstein refuses to provide his monster a mate; the genetically enhanced in *Gattaca* have a competitive advantage in romantic pursuits; and the citizens in Huxley's World State have had all reproductive powers removed from them, with pharmaceutically induced pregnancy simulations being the closest that any of them will experience to physiological reproduction. Power is also exerted through maximizing the efficiency or docility of the human body itself, demonstrating what Foucault terms the *anatomo-politics of the human body*.[14] In *Frankenstein*, the human physique is modified for strength; in *Gattaca*, the human genome is modified for health; and in *Brave New World*, both the human body and psyche are modified for passive contentedness.[15]

The targeted manipulation of the human form becomes the locus of a conservative resistance against the technoscience that made such manipulation possible. As such, I argue that these texts themselves and their application in bioethical debate are not only *politically motivated*, but also form part of the biopolitical discourse surrounding the integration of technology into existing and possible future political power structures. I will now explore the intersection of utopian sf studies, bioethics, and biopolitics in further detail by addressing a series of questions that illustrate how important this relationship can be, and how it may impact the future of reproductive biotechnology.

Who Is Using Science-Fictional References in Bioethical Debates?

Journalists and Online Bloggers

For this discussion it is necessary to first determine who is using sf imagery and language to describe biotechnology and bioethical concerns. One source is perhaps not surprising. Journalists wishing to sensationalize their articles on scientific advancements often invoke images of Frankensteinian monsters and dystopian futures, exploiting the anxieties of the general public for popularity and shock value.[16] Through a process known as "valenced framing," journalists can represent scientific developments through media coverage in a way deliberately intended to evaluate their associated political issues in either a positive or negative light.[17] Such framing has been shown to impact the strength of media consumers' attitudes of support or opposition, and is often used as a tactic in election campaigns to rally support for a preferred candidate or consolidate feelings of negativity toward an opponent.[18] What follows are some examples of mainstream media coverage regarding reproductive biotechnologies and how these were reframed by news aggregators and on social media blogs.

In September 1996, scientists in Tokyo announced that they had successfully incubated a goat fetus in an artificial womb for up to three weeks. *The New York Times* gave the article the title "The Artificial Womb is Born" and opened the piece with a quote from Huxley's *Brave New World*.[19] Without differentiating between animal and human fetuses, the chosen quote also included reference to the "Social Predestination Room," the site of government control over the population of Huxley's World State. Bearing no relevance to an artificially gestated goat fetus, it can only be assumed that the reference was intended to be associated with later comments in this news article predicting the generalizability of the extrauterine fetal incubation (EUFI) technique to humans. *ConsciousNewsMedia* republished this story in July 2013 from the *New York Times* archives, extending the title to read: "The Artificial Womb is Born and the World of the Matrix Begins," thus super-imposing the horrors of wide-scale human enslavement, destructive artificial intelligences, and ecological apocalypse onto the image of the goat fetus growing in a tank.[20] As there is no inherent connection between EUFI and social oppression by either a totalitarian government or a race of sentient machines, these examples demonstrate both sensationalism and politically motivated valence framing, intent on strengthening oppositional attitudes and persuading consumers to abandon even tacit support of EUFI and other ectogenesis technologies.

With regard to cloning, the UK's *Daily Mail* had this shocking headline to offer on January 20, 2013: "Wanted: 'Adventurous Woman' to Give Birth to Neanderthal Man—Harvard Professor Seeks Mother for Cloned Cave Baby."[21] While reminding readers that reproductive human cloning remains a criminal offense in Britain, the journalists claimed that the creation of a cloned neo-Neanderthal might be "so cutting-edge that it may not be covered by existing laws."[22] The implication is clear: such laws need immediate attention. To illustrate the apparent threat of recreating this genus of hominids, journalists Allan Hall and Fiona Macrae claimed the research was "reminiscent of *Jurassic Park*," a 1990s science-fictional book-cum-film in which cloned prehistoric animals turn on their human keepers.[23] Despite publishing a statement on January 22 by Harvard geneticist Professor George Church clearly articulating that he was *not* in fact seeking a surrogate to bear a Neanderthal clone, *The Huffington Post* nevertheless rapidly produced a poll in collaboration with global market research company YouGov gauging public opinion toward Neanderthal cloning. The leading online news aggregator then released an article on January 30 titled: "Neanderthal Clone Poll Finds Most Americans Oppose Cloning Human Relative."[24] As it turns out, the judgment that "most" Americans opposed this area of research was extrapolated from a poll of just 1000 adults, conducted over a two-day period beginning January 25.[25] Exaggeration and overly negative reporting may be serving to prejudice public opinion here regarding reproductive biotechnologies—in this instance, cloning.

Lawyers and Policy Makers

With the media frequently using such sf references as Huxley's World State and Shelley's monstrous creation to describe emerging technologies, it is not surprising to find members of the public adopting these same narratives when discussing the issues exposed to them through such media. However, as Tranter notes in his article "The Speculative Jurisdiction: The Science Fictionality of Law and Technology," lawyers and policy makers are also relying on sf references to debate the legal and bioethical issues surrounding possible future technologies. Paying particular attention to the issue of in vitro fertilization (IVF), Tranter argues that lawyers in the 1970s and 1980s often treated this technology as a threatening "next step" down the artificial human reproduction continuum, a progression they believed "would end with cloning and ectogenesis."[26] Since the creation of the world's first "test-tube" baby, Louise Brown, had been prefigured in sf, Tranter claims "it was almost mandatory for publishing on IVF and law to reference directly Huxley's *Brave New World* (1932) and Orwell's *Nineteen Eighty-Four* (1949)."[27] It is noteworthy that both these texts are well-known and distinctly dystopian. Tranter claims such "selective referencing" from within the sf genre has the potential to prejudice lawyers, policy makers, and the public against certain biotechnologies, by failing to appreciate the complexity of the issues being discussed.[28] Again, the association between Louise Brown's birth in 1978 and the rise of either of these dystopian futures is neither obvious nor justified. However, the use of Huxleyan and Orwellian imagery can be a persuasive tool influencing legal opinion toward reproductive biotechnologies.

When considering the ramifications of ectogenesis technologies on the legality of abortion for the *Arizona Law Review*, Robert J. Favole identifies a lack of detailed scientific writing on the topic to assist lawyers and policy makers in decision making. Referencing Laurence E. Karp's book *Genetic Engineering: Threat or Promise*? Favole agrees that this lack of available research is the result of "hostile attitudes" toward ectogenesis in the general public brought on by repeated association with Huxley's *Brave New World*.[29] In addition to such specific utopian sf textual references in legal scholarship, there are also general references to sf as a genre that do not single out individual texts. In her discussion of the legal status of cryopreserved embryos for the *Cardoza Journal of Law and Gender*, Bridget M. Fuselier outlines a scenario in which a daughter inherits her mother's frozen embryos and intends to gestate her genetic siblings against their father's wishes. Noting the lack of legal direction currently available for such a case, Fuselier claims that such a "science-fiction soap opera" demonstrates the need to establish policies regarding legal rights over un-implanted embryos.[30] Similarly, Glenn Cohen for the *Southern California Law Review* labels his "bathtub case," in which a person is unknowingly made a genetic parent after their skin cells have been harvested from a hotel bathtub, as a "science fiction thought experiment" designed to illustrate the need to clearly delineate the level of voluntariness required in a procreative act for the law to impose legal parenthood on a genetic parent.[31] In both of these instances, the authors go on to warn readers that such speculative fictions are fast becoming possible realities. Like Tranter, I argue that using sf references as a fear tactic in this way in legal scholarship exclusively serves socially conservative, technophobic political agendas, thereby preventing sf from living up to its potential to facilitate balanced public debates concerning these areas of policy and law.

Scientists and Researchers

One of the most interesting discoveries when exploring who is using sf references in literature concerning new reproductive biotechnologies is the extent to which scientists themselves use the language of sf to communicate their research. While many accuse journalists of distorting the scientific facts in their comparisons between emerging technologies and popular sf imagery, Sheryl N. Hamilton notes that following the successful cloning of Dolly the sheep in 1996, both her creator, Dr. Ian Wilmut, and leading bioethicist and biologist, Lee M.

Silver, released statements containing references to sf. With Wilmut defending his position that "cloning people should be in the realm of science fiction" and Silver declaring that now "all of science fiction is true," what Hamilton finds most revealing about these examples, like several of the law examples provided above, is the lack of specificity regarding sf:

> Why, however, would both these figures so prominently invoke science fiction? In making their claims, these scientists are doing something that at first glance might seem unlikely: they are directly linking biotechnological science with science fiction. Significantly, neither is referencing a particular sf novel, film or television program; both evoke the genre as a broader cultural trope. This rhetorical tactic seems curious in that science fiction might not necessarily be the best way to promote the credibility of science, particularly biotechnology, which commentators have noted has serious public relations problems.[32]

In his statement against human cloning, Wilmut is reaffirming that certain scientific ideas rightly belong *solely* in the fictional utopias of the scientific imaginary. Meanwhile, Silver is using the public's familiarity with sf to suggest that the advent of a cloned sheep demonstrates that the boundary between fact and fiction has collapsed—in short, if a human mind can conceive of a given technology, science can now make it a reality. While this is clearly an overstatement, whether a reader is likely to fear or take comfort from such a pronouncement will depend on the types of sf they have been exposed to, with the most thoroughly culturally engrained examples often tending toward dystopia and horror.

At the time these statements were released, exposure to at least one sf narrative was practically a given. When discussing Dolly, *Brave New World* references abounded both in the popular media and in bioethics literature, with both Dr. Frankenstein and his monster also making numerous appearances. According to Jane Maienschein, what became "very clear very quickly" following the entrance of Dolly into the cloning debate "was the importance of words and perspective."[33] She claims that the initial response of the US Congress to the possibilities Dolly represented was a "frightened reaction to the idea of runaway Frankenstein-like scientists copying people in their labs."[34] Again, the pervasiveness of sf language and imagery from *Brave New World* and *Frankenstein* in the debates surrounding cloning demonstrates the relationship between fictional, legal, political, and philosophical speculation.

Bioethicists

As noted above, the bioethical discourse regarding future reproductive biotechnologies incorporates the voices of journalists, scientists, lawyers, policy makers, and members of the public, all of whom may use sf as a common language to explore the social and political ramifications of medical and scientific development. However, the major focus of this essay is on the use of sf references by bioethicists, particularly in relation to the biopolitical significance of such references in bioethics scholarship. Some notable figures that choose to frame their arguments regarding future technologies with sf references include bioethicists Leon Kass, Ronald M. Green, and Nicholas Agar—the latter's seminal work *Liberal Eugenics* containing one reference to *Frankenstein*, nine to *Gattaca*, and the phrase "Brave New World" at least three times.[35] While certainly not a practice adopted by all bioethicists, such examples highlight a point of intersection between bioethics and popular culture.

Writing on scientism versus Luddism for the *American Journal of Bioethics*, Nancy C. Jones equates Luddism with the fear that scientific advancements will lead to "a reincarnated Frankenstein, deviously plotting to take over the world and create Aldous Huxley's *Brave New World*."[36] This quotation demonstrates the practice Hamilton observed as drawing on sf as a "broader cultural trope," the relevance of the vague references to two unrelated sf narratives being neither explicitly stated nor concerned with any degree of textual fidelity. It is significant that sf tropes, and sf as a trope, have become points of shared cultural history in bioethical debates, demonstrating the potential for sf to fill an important role in mediating between different stakeholders in our biotechnological future. What follows is a closer analysis of the use of utopian sf in bioethics, addressing each of the chosen texts in turn.

What Is Being Said Using Science-Fictional References in Bioethical Debates?

Brave New World

Huxley's cautionary tale is often referenced as a means of exploring the potential social impact of emerging reproductive biotechnologies on traditional views of sex, reproduction, and family.[37] However, many writers using *Brave New World* do not directly quote from the text, but rather rely on generalized conceptions of the text and its meaning. Used as an argument against both cloning and ectogenesis, a closer textual analysis of *Brave New World* reveals that the core of the dystopia is one of government-mandated social stratification and conditioning that owes very little to either of these technologies.

In their review of the ethical implications of cloning and stem cell research, Ted Peters and Gaymon Bennet Jr. identify three competing bioethical frameworks used to discuss these issues, one of which is simply labeled the "anti-Brave New World."[38] They claim that by invoking dystopian images of a Huxleyan future, proponents of this philosophy successfully prejudice the public against a variety of biotechnological developments, many of which could yield significant health benefits for the population. This is certainly the approach taken by Kass, whose extensive use of sf references to argue matters of ethical importance has led to some criticism from rival philosophers. S. Philip Morgan, Suzanne Shanahan, and Whitney Welsh claim that Kass's use of *Brave New World* as representative of the inevitable future of a society that permits human cloning, while certainly managing to frighten his readers, bears little resemblance to Huxley's actual text.[39] What is of interest here is not the level of textual fidelity, or lack thereof, found in these arguments, but rather their common assumption that all of their readers will at least be vaguely familiar with the dystopian future that Huxley imagined. That the language and imagery of *Brave New World* has infiltrated the bioethical literature on cloning to such an extent that an entire ethical framework can be reduced to "anti-Brave New World" is evidence that utopian sf and bioethics have become firmly entwined.

While praised for being accessible to a wide audience, Green's reliance on utopian sf in his argument against allowing "babies by design" is also criticized for an overdependence on negative portrayals of cloning and genetic engineering in sf texts, such as *Brave New World* and *Gattaca*.[40] Dena S. Davis claims that by using these dystopian projections of the future as a frame of reference for discussing current and advancing medical technologies, Green allows popular culture to paint these developments in a negative light, rather than providing persuasive ethical arguments to oppose them.[41] It is important to note that most of the sf texts appearing in bioethics literature depict technological change negatively, with positive counterexamples often lacking. Hilary S. Crew openly questions how different social attitudes toward cloning might be, were there another text as widely known as *Brave New World* that speculated on the potential benefits such technology could bring society.[42]

Association with this classic dystopia currently conflates the issues surrounding cloning and ectogenesis with those of the biopsychosocial totalitarianism of Huxley's World State, with its prenatally conditioned, drug-addled population and the total abolition of the traditional family unit. I argue that what can be seen now is a kind of feedback loop, in which *Brave New World*'s popularity was originally owed to the fact it spoke to contemporary concerns regarding the future of human reproduction, but now, in turn, provides the very foundation of that concern in the present. Thus, while utopian sf has the potential to engender a sense of scientific and technological enchantment, its selective use in bioethics literature can be seen as perpetuating a certain skepticism and disenchantment toward science, and particularly reproductive biotechnology.[43] As Tranter notes, "dystopian science fiction can be seen as profoundly anti-technological," and dystopias account for the majority of the sf texts being used in both legal and bioethical debate.[44] Recognizing that there are many worthy arguments against certain biotechnological developments, I maintain that it is important that the use of familiar sf tropes does not serve to unduly prejudice policy makers and the public against these technologies, and that the debate does not become one-sided.

Frankenstein

With regards to genetic engineering, Anne Murcott bemoans the repeated use of the term "Frankenfoods" to describe genetically modified crops, claiming that public opinion has been manipulated so artfully through its usage that "science fiction and science fact" can no longer be distinguished.[45] This is a view shared by many other stakeholders in the industry, with bioengineers protesting against accusations of hubris and irresponsibleness; the image of the scientist "playing God" being a common trope borrowed from sf. The catastrophic predictions for genetically engineered plants and animals are often attributed to sf sources, where the acceptance of such technologies at this early stage often represents a "slippery slope," which inevitably leads to genetic enhancements for humans (Franken-babies!),[46] the development of a genetic underclass (as in the Gattacan society),[47] or to ecological apocalypse (killer corn!).[48]

According to G. Pascal Zachary, the persistent image of scientific creations developing a "mind of their own" and bringing about the downfall of their creators "owes a lot to the image of Frankenstein's monster."[49] For obvious reasons, Shelley's original text contains no reference to genetic engineering, cloning, or stem cell research; however these are the bioethical and biopolitical debates in which Dr. Frankenstein and his creature are most frequently found today. Leaving aside the common misappropriation of the name of Shelley's protagonist to represent a scientifically manipulated product (which in the text is the unnamed monster), I believe the presence of references to *Frankenstein* in these debates speaks to the adaptability of Shelley's work and the timelessness of societal concerns regarding the risks of "playing God" through boundary-free science. More specifically, the text can be seen to warn against handing over the power to direct humanity's biological future to a subset of humans, whether scientists or state officials.

I argue that associating any future reproductive biotechnology with the stereotypical "mad scientist" Dr. Frankenstein automatically infers a moral judgment against that technology and those who would seek to control it. Chris Mason reminds us that Dr. Frankenstein's "transgressions into anthropoeia" cost him greatly, and that the moral of the story is clearly focused on respecting the boundaries of nature and science.[50] Given the cultural impact of Shelley's work, the term "Frankensteinian" when applied to any branch of science must necessarily carry with it strong connotations of moral disapprobation. Thus, when discussing scientific creations that "might out-Frankenstein Frankenstein," bioethicists are not critically engaging with Shelley's novel, but rather capitalizing on its cultural currency to support a particular sociopolitical argument, usually one opposed to a given area of scientific research.[51]

Gattaca

David A. Kirby claims that *Gattaca* is "an example of an extrapolative science fiction" in which the "filmmakers act as bioethicists."[52] He notes the strong influence sf can have on public perceptions of scientific development, explaining why *Gattaca* features so prominently in debates about the ethics of genetic susceptibility testing in the workplace, the risks of genetic discrimination, and the flaws inherent in genetic determinism. Kirby claims that sf "provides scholars a gauge of social concerns, social attitudes, and social change regarding science and technology."[53] Having always been intimately connected with scientific development, sf provides a forum to discuss the potential problems that future technologies may bring for human societies.[54] Lee Easton further argues that as sf is "neither simply social commentary nor precise scientific prediction," films like *Gattaca* can become part of the discourse determining our cultural response to new technologies.[55] By dramatizing ethical dilemmas rather than lecturing about them, sf can both educate and entertain, drawing on abstracted philosophical arguments to achieve real-world sociopolitical goals.[56]

In "Lifting the Genetic Veil of Ignorance: Is There Anything Really Unjust about Gattacan Society?" Sandra Shapshay uses *Gattaca* to illustrate the potential threat posed by unregulated genetic technologies on equality of opportunity in a near-future American society. Opposing those critics who claim the warnings in *Gattaca* are "overblown," Shapshay continues by discussing what she finds the most insidious aspect of *Gattaca*: the unintentional creation of a eugenicist society through the release of genetic enhancement technologies as personal services that couples "can take or leave" as they desire.[57] When invoked in bioethics literature, the name *Gattaca* alone can serve as a warning against the dangers of developing a society that is dependent on a legalized "genetic supermarket."[58] Unlike in *Brave New World*, *Gattaca* does not rely on state interference and control over human reproduction to produce a dystopia; in fact, the noninterference of the state can be seen as part of the problem, as it tolerates the perpetuation of unequal employment opportunities for those citizens existing outside the genetically enhanced bourgeoisie. According to Silver's logic, this makes *Gattaca* a more realistic image of the future:

> While Huxley guessed right about the power we would gain over the process of reproduction, I think he was dead wrong when it came to predicting *who* would use the power and for what purposes. What Huxley failed to understand, or refused to accept, was the driving force behind babymaking. It is individuals and couples... *not governments*—who will seize control of these new technologies. They will use some to reach otherwise unattainable reproductive goals and others to help their children achieve health, happiness, and success.[59]

In Niccol's text the ethical implications of allowing directed human evolution and eugenics to be practiced freely in society are well described, with *Gattaca* potentially representing one of the only frames of reference that many people have for considering the social and political implications of genetic technology. By encouraging sympathy for the unenhanced "in-Valids" of the Gattacan society, viewers can become politicized on the issue of genetic discrimination, a topic they may not have previously had any interest in or knowledge about. In this case resistance is not framed as rebellion against a powerful government, but rather as a duty to oppose the development of a technology that, if made available to individuals and couples, would serve to perpetuate social inequality.

Why Use Science-Fictional References in Bioethics and Biopolitics?

Utopian sf as Warning

There are many reasons why utopian sf lends itself to discussions on the nature of science and technology, and various motivations for employing the language of sf in bioethical and biopolitical debate. As the examples above demonstrate, whether or not someone has personally read *Brave New World*, *Frankenstein*, or watched *Gattaca*, there is an assumption when using these texts in debate that most members of the public will have a basic knowledge of the central issues arising in these texts. These include: warnings against trusting the hubristic and fallible scientist or the seemingly benevolent state authorities; the threat of unleashing the monstrously ugly; and the potential horrors of interfering with nature. Taking the example of *Frankenstein*, almost two centuries since its creation Shelley's horror sf classic still has resonance when applied to bioethical discussion, associating certain research projects with the potential for grotesque and usually unforeseen side effects. Lewis Wolpert claims that Shelley would be both "proud and shocked" at the impact her novel continues to have, both as a means of approaching concerns regarding scientific progress, and for its ability to condemn entire research proposals through the mere mention of its name. Here it is the popular conceptualization of *Frankenstein*, and not any in-depth analysis of the text, that is substituting for informed philosophical debate. In this way the characters in *Frankenstein* transcend their original text and association with them can automatically force one side of a bioethical debate onto the defensive.

Utopian sf as Shared Language Providing Metaphors and Names

Iina Hellsten claims that sf tropes can serve as metaphors for discussing science and technology, focusing the public's attention on specific areas of bioethical concern in a way that supports particular political agendas.[60] She notes that scientific discoveries are often framed in public debate as either "sensational breakthroughs and innovations," or as "steps toward creating new Frankensteins," claiming the latter argument deliberately uses this sf imagery to "block some aspects of the issue out of focus."[61] Since research into some potentially beneficial reproductive biotechnologies, such as cloning and ectogenesis, is currently subject to blanket bans in many countries, there is a need to establish a justification for banning scientific development in these areas that speaks to common public concerns. For this to occur, the public needs to be able to understand the ramifications of scientific advancement on society and the biopolitical implications of increased state and private interference into biological life. Maienschein notes that when engaging members of the public in discussions concerning policy regulation of scientific endeavors, a "familiar and conventionally accepted language" must be available.[62] This is where I believe sf can assist communication, by providing a shared language to aid the dissemination of scientific knowledge and political opinion. As Vint argues, "understanding the speculative discourses of biopolitics is imperative, and sf is in a privileged position to help us think through its anxieties and contradictions."[63]

As addressed earlier, bioethics represents the point of intersection between science and the humanities, scientists and the public. It is interesting that popular culture references have become part of the common language for such encounters, especially considering that the language of philosophy at large is not generally deemed to be particularly accessible. However, while some areas of philosophy deal in more abstract terms, I argue that the issues being decided upon in bioethical discussion can have real-world and immediate impact on all citizens, thus demonstrating the need to involve the general public in these discussions. Hamilton claims that sf provides an opportunity to reflect on science, and that the "persistent presence" of sf references in literature discussing emerging biotechnologies is "highly significant in terms of how cultural meanings of biotechnologies are constructed."[64] Susan M. Squier and Heather Latimer both go further to argue that fictional representations of reproductive biotechnologies are actually creating the social and cultural conditions for the posthuman.[65] Utopian sf achieves this goal by abandoning the constraints of the physical world and exploring alternative realities in which the posthuman already exists. Thus, the power of fiction in bioethical debates stems from writers being able to make issues come alive through the creation of imaginary characters that appeal to readers' sympathies—putting a name to the posthuman face, so to speak.[66] It has been well documented that the specific name given to an entity has a significant impact on how that entity is treated,[67] which explains how referring to future clones as "Brave New Worlders,"[68] and applying terms like "Franken-babies" to genetically engineered offspring can serve to manipulate public opinion regarding these technologies.[69]

Utopian sf as Bioethical and Political Instruction

Utopian sf references can also serve as educational tools in bioethics and biopolitics. For example, if asked to explain what genetic discrimination is, many among the uninitiated would not be able to come up with a definition, let alone an argument. However, when asked to elucidate the problem depicted in the Gattacan society, these same people can identify the elements of genetic discrimination as they affect Vincent, *Gattaca*'s unenhanced protagonist. Hence, Silver advises all geneticists to watch *Gattaca*, "if for no other reason than to understand the perception... of the public-at-large" toward genetic engineering.[70] This film also demonstrates that, although use of enhancing technologies in the future may be voluntary, access will still most likely be restricted to those financially able to pay for them, dramatizing the ethical dilemma of allowing a society of genetic "haves" and "have-nots" to develop. Françoise Baylis and Jason Scott Robert claim that genetic engineering will inevitably lead to genetic enhancements, "because so many of us are crass capitalists, eager to embrace biocapitalism."[71] These authors further argue that genetic enhancements "will never be available on demand," and thus have the potential to shatter the "myth of equality of opportunity" that many cultures rely on to secure social cooperation among citizens.[72] This is due to the fact that instead of promoting the idea that anyone can achieve success through hard work, genetic technologies will mean that the socially privileged will be born with enhancements that place the unenhanced at an even greater disadvantage in competition. *Gattaca* is thus a call-to-arms for policy makers and the public to fight against this kind of future.[73] Furthermore, the film explores the question of genuine choice regarding accessing reproductive biotechnologies, by suggesting that once a technology is available there will automatically be social pressure to use it. Take, for example, how IVF has influenced societal attitudes toward infertility, and it is clear that scientific advancement has the potential to alter our perceptions of the biological "norm." As is currently the case for IVF, the sociopolitical motivations governing the provision of future reproductive biotechnology services, including genetic therapies, will also warrant close examination by all stakeholders.

Utopian sf as "Worlds of Reference" to Explore

Finally, I believe the use of utopian sf references in bioethics literature is so prominent because the genre allows for the discussion of ethical concerns in an ostensibly non-biased hypothetical environment. In their article "Medical Ethics through the Star Trek Lens," James J. Hughes and John D. Lantos note that by creating a fictional reality that distances audiences from the real world, moral and ethical dilemmas can be illustrated in a way that allows "generalizability without insisting upon it."[74] They compare this role to that which is performed by the "philosophical parable," claiming that sf is able to analyze the same issues as those addressed by the hypothetical "mental exercises" of philosophical debate, but in less abstract terms.[75] The significant impact that sf has on public opinion earns the genre the label of sociopolitical "thought experiments"—artificial scenarios with real-world pedagogic value. Through the debates surrounding reproductive biotechnologies there are often repeated reminders that the possibilities depicted in sf are fast becoming realities, and that the public need to be aware of the potential risks of certain scientific advancements. The media frenzy often accompanying major developments in biotechnology demonstrates the role sf can play in public education, as many sf texts have already thoroughly explored the possible sociopolitical ramifications of emerging technologies well before the creation of a successful prototype. Taking the example of *Brave New World*, it was before the widespread acceptance of IVF that Huxley's future prediction dominated the literature on this topic, as there were no real-world examples of human societies that used this technology that could be pointed to. Now that we have evidence that IVF babies pose no immediate threat to democracy or human liberty, *Brave New World*'s use has mostly shifted to the cloning and ectogenesis debates, maintaining this element of *future* speculation. John D. Biggers notes that between the first successful attempt to culture preimplantation mammalian embryos in 1913, and the first public lecture by geneticist and biochemist J. B. S. Haldane on the possibility for in vitro reproductive technologies to lead to artificial gestation in 1923, discussions of ectogenesis were almost entirely contained within the realm of utopian sf.[76] Furthermore, Biggers claims that the idea of ectogenesis only became "widely disseminated" through the publication of *Brave New World* in 1932, thereby introducing this concept to the public.[77] Even today, the fictional worlds of sf still represent the only place where human ectogenesis currently exists.

With regards to the cloning debate, Patrick D. Hopkins claims that there is a "Brave New Rhetoric" in bioethical literature that is dependent not on Huxley's actual book but on the "trope connected with the book."[78] Vague references to the dystopia depicted in *Brave New World* are even used to stand in for philosophical argument entirely by those attempting to expose the potential risks of allowing the state to exercise political governance over human reproductive biology through the application of science. As Peter Firchow notes:

> So deep a mark has this work left on the modern literate mind that the mere mention of it evokes a whole complex of hostile attitudes towards science. It has become a kind of byword for a society in which the values (or nonvalues) of scientific technology are dominant, and which has therefore reduced man to a species of machines.[79]

It is difficult to imagine what the world might look like with the integration of the posthuman into human society. Cloning and ectogenesis, for example, carry with them a challenge to traditional views of parenthood and family, with Huxley's fictional society providing one possible scenario in which these challenges have essentially obliterated familial affection. Ethical concerns regarding reproductive biotechnologies often focus on the threat such techniques pose to the concept of the biological family unit, while also addressing the issue of the commodification of human life, and the predicted loss of individuality, autonomy, and dignity that this would entail. The inhabitants of Huxley's dystopia allow for the personification of these bioethical issues in the context of a totalitarian state. This illustrates that although technology itself may be politically neutral, its application and regulation rarely are. This lesson can be very effectively taught through exploring either *Brave New World*'s government control of human reproduction or *Gattaca*'s social policing of directed evolution.

Lawyers and bioethicists both lament the fact that the legal and ethical implications of new reproductive biotechnologies are often not sufficiently explored before they are released, with some claiming that they emerge in a moral and legal vacuum.[80] I argue that sf speculation can bridge this gap by exploring the sociopolitical issues facing possible future societies, particularly those encountering technologies with the potential to radically alter perceptions of the nature of humanity and definitions of human reproduction. Thus, the bioethical issues associated with new scientific discoveries can provide narrative fodder for sf, while the characters of utopian sf provide analogies for bioethicists for future people living in a world in which currently emerging biotechnologies are fully realized.

So Where Do We Go from Here?

There is an opportunity for utopian sf to continue to play a major role in shaping public education and opinion on bioethical matters. However, I argue that there is also a responsibility to provide alternatives in bioethics literature for the purely dystopian images of scientific advancement so often seen in texts like *Frankenstein, Brave New World,* and *Gattaca*. Not all uses of utopian sf in bioethics need to be negatively motivated, as these speculative fictions also have the potential to promote accessible public debate concerning the possible benefits of emerging reproductive biotechnologies. For example, Lois McMaster Bujold's *Vorkosigan Saga*, set in a world in which IVF and ectogenesis have replaced biological gestation, represents a mostly positive alternative to Huxley's dystopia. While references to this text can be found within the realm of disability studies, due to the presence of a disabled protagonist, it is almost entirely absent from the bioethical literature concerning ectogenesis. Bujold's complex treatment of reproductive biotechnologies and their associated ethical and political concerns is more balanced than Huxley's, demonstrating both the pros and cons of technological intervention in human reproduction.

H. Bruce Franklin defines sf as "the literature which, growing with science and technology, evaluates it and relates is meaningfully to the rest of human existence."[81] This highlights the future of sf as a medium for the negotiation of science into society, bridging the two "gaps" that both Istvan Csicery-Ronay and Eugene Thacker identify: the "technical gap" between what can be conceived in the imagination and that which is currently available, and the "ethico-political gap" between what *could* become possible, and what we *should* allow to occur in the future.[82] This relies on sf continuing to anticipate future possible worlds, in which issues like posthuman citizenship, clone rights, and the reproductive liberty of the genetically enhanced are being actively debated. Just as Fredric Jameson defends the special relationship between sf and social development,[83] Csicery-Ronay believes sf serves as a "mode of awareness" of the scientific world, existing between the poles of imagination and realization of "scientific-technological transformations."[84] However, as discussed throughout this essay, within bioethics scholarship sf is currently being exploited almost exclusively to promote socially conservative political agendas.

Although most of the sf references used in bioethical literature involve negative portrayals of reproductive biotechnologies, positive examples do exist in the genre and could be used to argue in favor of technological development. Kirby notes that while in the real world human genetic engineering is most commonly employed to correct life-threatening genetic conditions, sf stories "rarely feature this as a plot element because it does not make for an action-packed narrative."[85] This demonstrates Jameson's concern about eutopian narratives, that they will always elicit the "stirrings of a fear... the fear of boredom."[86] However, while certainly underrepresented in the genre, there have been a number of successful sf stories that have characterized reproductive biotechnologies and alternative methods of procreation in a positive light. Marge Piercy's utopian sf, *Woman on the Edge of Time* (1976), depicts a fictional society in which ectogenesis has brought about total sexual equality.[87] Similarly, Ursula K. Le Guin's *The Left Hand of Darkness* (1969) explores an alien society in which gender distinctions do not exist, and thus the burdens of reproduction fall equally across all members of the population. The fact that these texts are comparatively overlooked in bioethical literature exposes the dominant conservatism behind the use of sf language and imagery in bioethical debate.

In her analysis of sf's potential to assist in comprehending "new biopolitical subjectivities," Squier claims sf is now "only a grim commentary on life as we must live it; no longer fantasy, but documentary."[88] However, I argue that utopian narratives are unique in that they serve as powerful persuasive tools shaping how we imagine and pursue possible futures. Sargent claims utopia "is at the root of all radicalism... the archetype and harbinger of social change," and that visions of the future can have a tangible effect on social and political realities.[89] Thus, utopian sf has both persuasive and subversive power, fostering an interest in "social questions" among readers, and providing a palatable form of political instruction.[90] Redressing the underutilization of sf references to also promote liberal, pro-technological political agendas, represents an opportunity for sf to be engaged in more balanced discussions of future reproductive biotechnologies.

Conclusion

Necessarily concerned with the sociopolitical aspects of an altered world, utopian sf yields a rich source of hypotheticals for bioethicists to explore. Ideas and images from utopian sf texts feature prominently in many bioethical debates, particularly those connected with the artificial creation or manipulation of human life. This demonstrates that sf authors may be in a unique position to influence public opinion in these areas, by providing a common language to mediate between scientists and the public, and producing fictional "worlds of reference" for bioethics to explore. As an exploration of the scientific imaginary, I argue that sf has a distinct role to play in biopolitics, but one that has only been fully exploited by one side of the technoscience political divide. As such, there exists an opportunity for engaging the public with more positive representations of future reproductive biotechnologies, including cloning, ectogenesis, and genetic engineering. I believe that the use of sf language in bioethics highlights an opportunity for sf to be involved in the re-enchantment of science and technology, while also bearing a social responsibility for providing meaningful philosophical and political debate on the potential impact of emerging technologies on society.

Notes

1. Darko Suvin, *Metamorphoses of Science Fiction: On the Poetics and History of a Literary Genre* (New Haven: Yale University Press, 1979), 61.
2. Ibid.; Darko Suvin, *Defined by a Hollow: Essays on Utopia, Science Fiction and Political Epistemology* (Bern: Peter Lang AG, 2010), 30.
3. Suvin, *Metamorphoses of Science Fiction*,3 8.
4. Lyman Tower Sargent, "Three Faces of Utopianism Revisited," *Utopian Studies* 5, no. 1 (1994): 9.
5. Ibid., 3.
6. Brian M. Stableford, and David Langford, "Utopias," in *The Encyclopedia of Science Fiction*, edited by John Clute, David Langford, Peter Nicholls, and Graham Sleight. London: Gollancz, updated 7 February 2014. Web. Accessed 11 February 2014. http://www.sf-encyclopedia.com/entry/utopias.
7. Kieran Tranter, "The Speculative Jurisdiction: The Science Fictionality of Law and Technology," *Griffith Law Review* 20, no. 4 (2011): 818.
8. Sherryl Vint, "Introduction: Science Fiction and Biopolitics," *Science Fiction Film and Television* 4, no. 2 (2011): 161.
9. Andrew Milner, *Literature, Culture and Society* 2nd edn. (London: Routledge, 2004), 137.
10. Mary Shelley, *Frankenstein: Or, the Modern Prometheus* (London: Lackington, Hughes, Harding, Mavor & Jones, 1818), vol. 3, chap. 3, pg. 1.
11. Some examples can be seen in the choice of titles for bioethical texts dealing with genetic engineering, such as Bernard E. Rollin's *The Frankenstein Syndrome: Ethical and Social Issues in the Genetic Engineering of Animals* (Cambridge: Cambridge University Press, 1995); and Byron L. Sherwin's "Golems in the Biotech Century," *Zygon* 42, no. 1 (2007): 133–144. Michael Mulkay notes that in the area of embryo research in particular, both the experiments and the scientists involved are often discussed "in Frankensteinian terms." He analyses examples from within philosophical, political, legal, and popular media publications, and the impact these have had on public opinion and legislation toward embryo research ("Frankenstein and the Debate Over Embryo Research," *Science, Technology, & Human Values* 21, no. 2 (1996): 161).
12. Sandra Shapshay, "Lifting the Genetic Veil of Ignorance: Is There Anything Really Unjust about Gattacan Society?" In Sandra Shapshay (ed.) *Bioethics at the Movies* (Baltimore: The John Hopkins University Press, 2009), 88.

13. Michel Foucault, *The History of Sexuality: An Introduction Volume I* (New York: Vintage Books, 1990), 139.
14. Ibid.
15. This fits Foucault's model, which states that *anatomo-politics of the human body* involves "its disciplining, the optimization of its capabilities, the extortion of its forces, the parallel increase of its usefulness and its docility, its integration into systems of efficient and economic controls." Ibid.
16. Anne Murcott, "Public Beliefs about GM Foods: More on the Makings of a Considered Sociology," *Medical Anthropology Quarterly* 15, no. 1 (2001): 15.
17. A. R. T. Schuck, and C. H. de Vreese, "Between Risk and Opportunity: News Framing and Its Effects on Public Support for EU Enlargement," *European Journal of Communication* 21, no. 1 (2006): 6.
18. George Y. Bizer, Jeff T. Larsen, and Richard E. Petty, "Exploring the Valence-Framing Effect: Negative Framing Enhances Attitude Strength," *Political Psychology* 32, no. 1 (2011): 65.
19. Perri Klass, "The Artificial Womb Is Born," *The New York Times* online, September 29, 1996, available at: http://www.nytimes.com/1996/09/29/magazine/the-artificial-womb-is-born.html?pagewanted=all&src=pm.
20. Anonymous, "The Artificial Womb Is Born and the World of the Matrix Begins," *Conscious News Media* online, July 11, 2013, http://consciousnewsmedia.blogspot.com.au/2013/11/the-artificial-womb-is-born-and-world.html#.UvAah3eSwVk.
21. Allan Hall, and Fiona Macrae, "Wanted: 'Adventurous Woman' to Give Birth to Neanderthal Man – Harvard Professor Seeks Mother for Cloned Cave Baby," *Daily Mail Online*, January 20, 2013, http://www.dailymail.co.uk/news/article-2265402/Adventurous-human-woman-wanted-birth-Neanderthal-man-Harvard-professor.html.
22. Ibid.
23. Ibid.
24. Emily Swanson, "Neanderthal Clone Poll Finds Most Americans Oppose Cloning Human Relative," *The Huffington Post*, January 30, 2013, http://www.huffingtonpost.com/2013/01/30/neanderthal-clone-poll_n_2585096.html.
25. Using only four questions, this poll asked interviewees whether they would support cloning woolly mammoths or Neanderthals, particularly if the latter required the use of a surrogate human mother, and finally whether they believed scientists should be allowed to attempt full human cloning. The progression ran on a sharp incline: animal cloning → near-human cloning → near-human cloning involving a "real" human parent → full human cloning. *The Huffington Post/YouGov Omnibus poll*, January 25–26, 2013, http://big.assets.huffingtonpost.com/toplines12613.pdf.
26. Tranter, "The Speculative Jurisdiction," 826.
27. Ibid., 827.
28. Ibid., 836.
29. Robert J. Favole, "Artificial Gestation: New Meaning for the Right to Terminate a Pregnancy," *Arizona Law Review* 21 (1979): 760.

30. Bridget M. Fuselier, "The Wisdom of Solomon: We Cannot Split the Pre-Embryo," *Cardozo Journal of Law and Gender* 17 (2010–11): 511.
31. Glenn Cohen, "The Right Not to Be a Genetic Parent?" *Southern California Law Review* 81 (2007–8): 1125.
32. Sheryl N. Hamilton, "Traces of the Future: Biotechnology, Science Fiction, and the Media," *Science Fiction Studies* 30, no. 2 (2003): 267.
33. Jane Maienschein, "Part II: What's in a Name: Embryos, Clones, and Stem Cells," *The American Journal of Bioethics* 2, no. 1 (2002): 16.
34. Ibid., 17.
35. Nicholas Agar, *Liberal Eugenics: In Defence of Human Enhancement* (Malden: Blackwell Publishing, 2004).
36. Nancy C. Jones, "Scientism or Luddism: Is Informed Ethical Possible?" *The American Journal of Bioethics* 4, no. 1 (2004): 18.
37. Chris Mason, "Making People: Today's Wariness of Reproductive Technologies Stems from Myths, Legends and Hollywood," *Nature* 471, no. 7338 (2011): 297.
38. Ted Peters and Gaymon Bennett, Jr., "Stem Cell Research and the Claim of the Other in the Human Subject," *Dialog: A Journal of Theology* 43, no. 3 (2004): 185.
39. S. Philip Morgan, Suzanne Shanahan, and Whitney Welsh, "Brave New Worlds: Philosophy, Politics, and Science in Human Biotechnology," *Population and Development Review* 31, no. 1 (2005): 131.
40. Ronald M. Green, *Babies by Design: The Ethics of Genetic Choice* (New Haven: Yale University Press, 2008).
41. Dena S. Davis, "From YUCK to WOW and Back Again: A Review Article," *Religious Studies Review* 35, no. 3 (2009): 147.
42. Hilary S. Crew, "Not so Brave a World: The Representation of Human Cloning in Science Fiction for Young Adults," *The Lion and the Unicorn* 28, no. 2 (2004): 204.
43. Hamilton, "Traces of the Future," 278.
44. Tranter, "The Speculative Jurisdiction," 838.
45. Murcott, "Public Beliefs about GM Foods," 15.
46. Heather Latimer, "Reproductive Technologies, Fetal Icons, and Genetic Freaks: Shelley Jackson's Patchwork Girl and the Limits and Possibilities of Donna Haraway's Cyborg," *Modern Fiction Studies* 57, no. 2 (2011): 325.
47. Shapshay, "Lifting the Genetic Veil of Ignorance," 88.
48. Samantha Seiple, and Todd Seiple, *Mutants, Clones and Killer Corn: Unlocking the Secrets of Biotechnology* (Minneapolis: Lerner Publications Company, 2005).
49. G. Pascal Zachary, "Ethics for a Very Small World," *Foreign Policy* 137 (2003): 45.
50. Mason, "Making People," 297.
51. Alfred Gellhorn, "Medical Ethics: So What's the Story?" *In Vitro* 13, no. 10 (1977): 592.
52. David A. Kirby, "The New Eugenics in Cinema: Genetic Determinism and Gene Therapy in 'GATTACA'," *Science Fiction Studies* 27, no. 2 (2000): 199.
53. David A. Kirby, "The Devil in Our DNA: A Brief History of Eugenics in Science Fiction Films," *Literature and Medicine* 26, no. 1 (2007): 84.

54. Louise Economides, "Recycled Creatures and Rogue Genomes: Biotechnology in Mary Shelley's *Frankenstein* and David Mitchell's *Cloud Atlas*," *Literature Compass* 6, no. 3 (2009): 615; Gerard Mannion, "Genetics and the Ethics of Community," *HeyJ* 47 (2006): 247.
55. Lee Easton, "Passing Genes in GATTACA, or Straight Genes for the Queer Guy." In Lee Easton and Randy Schroeder (eds.) *The Influence of Imagination: Essays on Science Fiction and Fantasy as Agents of Social Change* (Jefferson: McFarland & Company Inc, 2007), 70–71.
56. James J. Hughes, and John D. Lantos, "Medical Ethics through the Star Trek Lens," *Literature and Medicine* 20, no. 1 (2001): 32.
57. Shapshay, "Lifting the Genetic Veil of Ignorance," 88.
58. Curtis A. Kin, "Coming Soon to the 'Genetic Supermarket' Near You," *Stanford Law Review* 48, no. 6 (1996): 1573.
59. Lee M. Silver *Remaking Eden: Cloning, Genetic Engineering and the Future of Humankind?* (London: Phoenix Giant, 1998), 10.
60. Iina Hellsten, "Popular Metaphors of Biosciences: Bridges over Time?" *Configurations* 16, no. 1 (2008): 22.
61. Ibid., 16–17.
62. Maienschein, "Part II: What's in a Name," 15.
63. Vint, "Introduction," 161.
64. Hamilton, "Traces of the Future," 268–270.
65. Susan M. Squier, "Reproducing the Posthuman Body: Ectogenetic Fetus, Surrogate Mother, Pregnant Man." In Judith Halberstam, and Ira Livingston (eds.) *Posthuman Bodies* (Bloomington: Indiana University Press, 1995); Latimer, "Reproductive Technologies," 332.
66. Evelyn Tsitas, "The Role of the Creative Arts in Bioethical Debates," *Law and Justice Journal* 6, no. 2 (2006): 260.
67. Jennifer C. Lahl, "Caveat Emptor," *The American Journal of Bioethics* 4, no. 1 (2004): 20.
68. Nick Bostrom, "In Defence of Posthuman Dignity," *Bioethics* 19, no. 3 (2005): 206.
69. Latimer, "Reproductive Technologies," 332.
70. Lee M. Silver, "Genetics Goes to Hollywood," *Nature Genetics* 17 (1997): 260.
71. Françoise Baylis, and Jason Scott Robert, "The Inevitability of Genetic Enhancement Technologies," *Bioethics* 18, no. 1 (2004): 19.
72. Ibid., 15.
73. Ken McLeod, "GATTACA." In Lee Easton and Randy Schroeder (eds.) *The Influence of Imagination: Essays on Science Fiction and Fantasy as Agents of Social Change* (Jefferson: McFarland & Company Inc, 2007), 176.
74. Hughes and Lantos, "Medical Ethics through the Star Trek Lens," 32.
75. Ibid.
76. John D. Biggers, "IVF and Embryo Transfer: Historical Origin and Development," *Reproductive BioMedicine Online* 25 (2012): 119.
77. Ibid.

78. Patrick D. Hopkins, "Bad Copies: How Popular Media Represent Cloning as an Ethical Problem," *The Hastings Center Report* 28, no. 2 (1998): 11.
79. Peter Firchow, "Science and Conscience in Huxley's 'Brave New World'," *Contemporary Literature* 16, no. 3 (1975): 300.
80. John Edwards, "New Conceptions: Biosocial Innovations and the Family," *Journal of Marriage and Family* 53, no. 2 (1991): 358.
81. H. Bruce Franklin, *Future Perfect: American Science Fiction of the Nineteenth Century: An Anthology* (Oxford and New York: Oxford University Press, 1966), 1.
82. Istvan Csicery-Ronay, "The SF of Theory: Baudrillard and Haraway," *Science Fiction Studies* 18, no. 3 (1991): 387; Eugene Thacker, "SF, Technoscience, Net. art: The Politics of Extrapolation," *Art Journal* 59, no. 3 (2000): 66.
83. Fredric Jameson, "Progress versus Utopia; Or, Can We Imagine the Future?" *Science Fiction Studies* 9, no. 2 (1982): 147.
84. Csicery-Ronay, "The SF of Theory," 387.
85. Kirby, "The Devil in Our DNA," 97; Kirby, "The New Eugenics," 196.
86. Fredric Jameson, *Archaeologies of the Future: The Desire Called Utopia and Other Science Fiction* (London: Verso, 2005), 81.
87. Rosemarie Tong, *Feminist Thought: A Comprehensive Introduction* (Oxon: Routledge, 1997), 76.
88. Susan M. Squier, *Liminal Lives: Imagining the Human at the Frontiers of Biomedicine* (Durham: Duke University Press, 2004), 212.
89. Lyman Tower Sargent, "Authority and Utopia: Utopianism in Political Thought," *Polity Journal* 14, no. 4, (1982): 566.
90. Mulford Q. Sibley, "Apology for Utopia II: Utopia and Politics," *Journal of Politics* 2, no. 2 (1940): 186.

Bibliography

Agar, Nicholas. *Liberal Eugenics: In Defence of Human Enhancement.* Malden: Blackwell, 2004.

Baylis, Françoise, and Jason Scott Robert. "The Inevitability of Genetic Enhancement Technologies." *Bioethics* 18, no. 1 (2004): 1–26.

Biggers, John D. "IVF and Embryo Transfer: Historical Origin and Development." *Reproductive BioMedicine Online* 25 (2012): 118–127.

Bizer, George Y., Jeff T. Larsen, and Richard E. Petty. "Exploring the Valence-Framing Effect: Negative Framing Enhances Attitude Strength." *Political Psychology* 32, no. 1 (2011): 59–80.

Bostrom, Nick. "In Defence of Posthuman Dignity," *Bioethics* 19, no. 3 (2005): 202–214.

Cohen, Glenn. "The Right Not to Be a Genetic Parent?" *Southern California Law Review* 81 (2007–8): 1115–1196.

Conscious News Media. "The Artificial Womb Is Born and the World of the Matrix Begins." *ConsciousNewsMedia* online, July 11, 2013. http://consciousnewsmedia.

blogspot.com.au/2013/11/the-artificial-womb-is-born-and-world.html#. UvAah3eSwVk.

Crew, Hilary S. "Not So Brave a World: The Representation of Human Cloning in Science Fiction for Young Adults." *The Lion and the Unicorn* 28, no. 2 (2004): 203–221.

Csicery-Ronay, Istvan. "The SF of Theory: Baudrillard and Haraway." *Science Fiction Studies* 18, no. 3 (1991): 387–404.

Davis, Dena. "From YUCK to WOW and Back Again: A Review Article." *Religious Studies Review* 35, no. 3 (2009): 24–27.

Easton, Lee. "Passing Genes in GATTACA, or Straight Genes for the Queer Guy." In *The Influence of Imagination: Essays on Science Fiction and Fantasy as Agents of Social Change*, edited by Lee Easton and Randy Schroeder, 70–82. Jefferson: McFarland & Company Inc, 2007.

Economides, Louise. "Recycled Creatures and Rogue Genomes: Biotechnology in Mary Shelley's *Frankenstein* and David Mitchell's *Cloud Atlas*." *Literature Compass* 6, no. 3 (2009): 615–631.

Edwards, John. "New Conceptions: Biosocial Innovations and the Family." *Journal of Marriage and Family* 53, no. 2 (1991): 349–360.

Favole, Robert J. "Artificial Gestation: New Meaning for the Right to Terminate a Pregnancy." *Arizona Law Review* 21 (1979): 755–776.

Firchow, Peter. "Science and Conscience in Huxley's *Brave New World*." *Contemporary Literature* 16, no. 3 (1975): 301–316.

Foucault, Michel. *The History of Sexuality Volume I: An Introduction*. New York: Vintage Books, 1990.

Franklin, H. Bruce. *Future Perfect: American Science Fiction of the Nineteenth Century: An Anthology*. Oxford and New York: Oxford University Press, 1966.

Fuselier, Bridget M. "The Wisdom of Solomon: We Cannot Split the Pre-Embryo." *Cardozo Journal of Law and Gender* 17 (2010–11): 507–526.

Gellhorn, Alfred. "Medical Ethics: So What's the Story?" *In Vitro* 13, no. 10 (1977): 588–594.

Green, Ronald M. *Babies by Design: The Ethics of Genetic Choice*. New Haven: Yale University Press, 2008.

Hall, Allan, and Fiona Macrae. "Wanted: "Adventurous Woman" to Give Birth to Neanderthal Man – Harvard Professor Seeks Mother for Cloned Cave Baby." *Daily Mail Online*, January 20, 2013. http://www.dailymail.co.uk/news/article-2265402/Adventurous-human-woman-wanted-birth-Neanderthal-man-Harvard-professor.html.

Hamilton, Sheryl N. "Traces of the Future: Biotechnology, Science Fiction, and the Media." *Science Fiction Studies* 30, no. 2 (2003): 267–282.

Hellsten, Iina. "Popular Metaphors of Biosciences: Bridges Over Time?" *Configurations* 16, no. 1 (2008): 11–32.

Hopkins, Patrick D. "Bad Copies: How Popular Media Represent Cloning as an Ethical Problem." *The Hastings Center Report* 28, no. 2 (1998): 6–13.

Hughes, James J., and John D. Lantos. "Medical Ethics Through the Star Trek Lens." *Literature and Medicine* 20, no. 1 (2001): 26–38.

Huxley, Aldous. *Brave New World*. London: Chatto and Windus, 1932.
Jameson, Fredric. "Progress versus Utopia; Or, Can We Imagine the Future?" *Science Fiction Studies* 9, no. 2 (1982): 147–158.
———. *Archaeologies of the Future: The Desire Called Utopia and Other Science Fiction*. London: Verso, 2005.
Jones, Nancy C. "Scientism or Luddism: Is Informed Ethical Possible?" *The American Journal of Bioethics* 4, no. 1 (2004): 18–20.
Kin, Curtis A. "Coming Soon to the "Genetic Supermarket" Near You." *Stanford Law Review* 48, no. 6 (1996): 1573–1604.
Kirby, David A. "The New Eugenics in Cinema: Genetic Determinism and Gene Therapy in 'GATTACA.'" *Science Fiction Studies* 27, no. 2 (2000): 193–215.
———. "The Devil in Our DNA: A Brief History of Eugenics in Science Fiction Films." *Literature and Medicine* 26, no. 1 (2007): 83–108.
Klass, Perri. "The Artificial Womb Is Born," *The New York Times* online, September 29, 1996. http://www.nytimes.com/1996/09/29/magazine/the-artificial-womb-is-born.html?pagewanted=all&src=pm.
Lahl, Jennifer C. "Caveat Emptor." *The American Journal of Bioethics* 4, no. 1 (2004): 20–21.
Latimer, Heather. "Reproductive Technologies, Fetal Icons, and Genetic Freaks: Shelley Jackson's Patchwork Girl and the Limits and Possibilities of Donna Haraway's Cyborg." *Modern Fiction Studies* 57, no. 2 (2011): 318–335.
Maienschein, Jane. "Part II: What's in a Name: Embryos, Clones, and Stem Cells." *The American Journal of Bioethics* 2, no. 1 (2002): 12–19.
Mannion, Gerard. "Genetics and the Ethics of Community." *HeyJ* 47 (2006): 226–256.
Mason, Chris. "Making People: Today's Wariness of Reproductive Technologies Stems from Myths, Legends and Hollywood." *Nature* 471, no. 7338 (2011): 297–299.
McLeod, Ken. "GATTACA." In *The Influence of Imagination: Essays on Science Fiction and Fantasy as Agents of Social Change*, edited by in Lee Easton and Randy Schroeder, 174–176. Jefferson: McFarland & Company Inc, 2007.
Milner, Andrew. *Literature, Culture and Society* 2nd ed. London: Routledge, 2004.
Morgan, S. Philip, Suzanne Shanahan, and Whitney Welsh. "Brave New Worlds: Philosophy, Politics, and Science in Human Biotechnology." *Population and Development Review* 31, no. 1 (2005): 127–141.
Mulkay, Michael. "Frankenstein and the Debate Over Embryo Research." *Science, Technology, & Human Values* 21, no. 2 (1996): 157–176.
Murcott, Anne. "Public beliefs about GM foods: More on the Makings of a Considered Sociology." *Medical Anthropology Quarterly* 15, no. 1 (2001): 9–19.
Niccol, Andrew. *Gattaca*. Culver City, CA: Columbia Pictures Corporation, 1997.
Peters, Ted and Gaymon Bennett, Jr. "Stem Cell Research and the Claim of the Other in the Human Subject." *Dialog: A Journal of Theology* 43, no. 3 (2004): 184–204.

Rollin, Bernard E. *The Frankenstein Syndrome: Ethical and Social Issues in the Genetic Engineering of Animals*. Cambridge: Cambridge University Press, 1995.

Sargent, Lyman Tower. "Authority and Utopia: Utopianism in Political Thought." *Polity Journal* 14, no. 4, (1982): 565–584.

———. "Three Faces of Utopianism Revisited," *Utopian Studies* 5, no. 1 (1994): 1–37.

Schuck, A. R. T. and C. H. de Vreese. "Between Risk and Opportunity: News Framing and Its Effects on Public Support for EU Enlargement." *European Journal of Communication* 21, no. 1 (2006): 5–32.

Seiple, Samantha, and Todd Seiple. *Mutants, Clones and Killer Corn: Unlocking the Secrets of Biotechnology*. Minneapolis: Lerner Publications Company, 2005.

Shapshay, Sandra. "Lifting the Genetic Veil of Ignorance: Is There Anything Really Unjust about Gattacan Society?" In *Bioethics at the Movies*, edited by in Sandra Shapshay, 87–101. Baltimore: The John Hopkins University Press, 2009.

Shelley, Mary. *Frankenstein: Or, the Modern Prometheus*. London: Lackington, Hughes, Harding, Mavor & Jones, 1818.

Sherwin, Byron L. "Golems in the Biotech Century." *Zygon* 42, no. 1 (2007): 133–144.

Sibley, Mulford Q. "Apology for Utopia II: Utopia and Politics." *Journal of Politics* 2, no. 2 (1940): 165–188.

Silver, Lee M. *Remaking Eden: Cloning, Genetic Engineering and the Future of Humankind?* London: Phoenix Giant, 1998.

———. "Genetics Goes to Hollywood." *Nature Genetics* 17 (1997): 260.

Squier, Susan M. "Reproducing the Posthuman Body: Ectogenetic Fetus, Surrogate Mother, Pregnant Man." In *Posthuman Bodies*, edited by in Judith Halberstam and Ira Livingston. Bloomington: Indiana University Press, 1995.

———. *Liminal Lives: Imagining the Human at the Frontiers of Biomedicine*. Durham: Duke University Press, (2004).

Stableford, Brian M., and David Langford. "Utopias." In *The Encyclopedia of Science Fiction*, edited by John Clute, David Langford, Peter Nicholls, and Graham Sleight. London: Gollancz, updated 7 February 2014. Web. Accessed 11 February 2014. http://www.sf-encyclopedia.com/entry/utopias.

Suvin, Darko. *Metamorphoses of Science Fiction: On the Poetics and History of a Literary Genre*. New Haven: Yale University Press, 1979.

———. *Defined by a Hollow: Essays on Utopia, Science Fiction and Political Epistemology*. Bern: Peter Lang AG, 2010.

Swanson, Emily. "Neanderthal Clone Poll Finds Most Americans Oppose Cloning Human Relative." *The Huffington Post*, January 30, 2013. http://www.huffingtonpost.com/2013/01/30/neanderthal-clone-poll_n_2585096.html.

Thacker, Eugene. "SF, Technoscience, Net.art: The Politics of Extrapolation." *Art Journal* 59, no. 3 (2000): 64–73.

Tong, Rosemarie. *Feminist Thought: A Comprehensive Introduction*. Oxon: Routledge, 1997.

Tranter, Kieran. "The Speculative Jurisdiction: The Science Fictionality of Law and Technology." *Griffith Law Review* 20, no. 4 (2011): 817–850.

Tsitas, Evelyn. "The Role of the Creative Arts in Bioethical Debates." *Law and Justice Journal* 6, no. 2 (2006): 255–265.

Vint, Sherryl. "Introduction: Science Fiction and Biopolitics." *Science Fiction Film and Television* 4, no. 2 (2011): 161–172.

Zachary, G. Pascal. "Ethics for a Very Small World." *Foreign Policy* 137 (2003): 108–109.

CHAPTER 5

Decolonizing the Future: Biopolitics, Ethics, and Foresight through the Lens of Science Fiction

Selena Middleton

Science fiction is a genre that straddles the divide between present and future. It is a genre that gathers the dust of present-day issues, concerns, and questions, molds them into human form and breathes into them a life that is compelled onward into a speculative future. While science fiction often bends the rule of the plausible, dipping a toe into the fantastic, the genre remains rooted in present systems and social structures. In this way, science fiction inspires a relationship with the future that necessarily questions the present structures that will inevitably influence it.

Bruno Latour highlights one important element in this relationship, referencing the iconic scene wherein Mary Godwin, Percy Bysshe Shelley, and Lord Byron engage in a contest to create the most chilling ghost story. This contest became the point of origin of the story that would become Mary Shelley's *Frankenstein*. This story would, in turn, become a kind of creation mythology in humankind's fledgling relationship with technology. *Frankenstein* set the tone for a continuing fear of and nonengagement with the products of scientific knowledge. In addressing the human response to future technologies—in this case, an immortalizing biotechnology—Mary Shelley tapped into a profound fear of technology that confounds humanity's relationship with nature in Dr. Frankenstein's response to his own creation. Latour points out that

> Dr. Frankenstein's crime was not that he invented a creature through some combination of hubris and high technology, but rather that he *abandoned the creature to itself.* When Dr. Frankenstein meets his creation on a glacier in the Alps, the monster claims that it was not *born* a monster, but that it became a criminal only *after* being left alone by his horrified creator, who fled the laboratory once the horrible thing twitched to life. [...] *Frankenstein* foresees that the gigantic sins that were to be committed would hide a much greater sin. It is not the case that we have failed to care for Creation, but that we have failed to care for our technological creations. We confuse the monster for its creator and blame our sins against Nature upon our creations. But our sin is not that we created technologies but that we failed to love and care for them. It is as if we decided that we were unable to follow through with the education of our children.[1]

If our technologies will eventually make their way in the world and their very existence leads to the creation of new technologies that will also exist in and affect the world we live in, then the creators of these technologies certainly have a responsibility toward their "children." That responsibility extends not only into the future—demanding an ongoing relationship with the technology—but also into the past (or, as we are thinking of the impact of future technologies, the present). What should we have been doing to ensure a functional relationship with our creations? What should we be doing now? These questions suggest the very problems with which science fiction is preoccupied; namely, that while we may not be able to predict the technological advances that the future will bring, we can prepare ourselves and our social structures to receive them.

Relevant to the idea of negotiating between the present and future is the current relationship (or lack of relationship) between art and science. Latour asserts that "Science is the shibboleth that defines the right direction of the arrow of time," because science is able to deftly define the two oppositional forces of human progress, the "morass of ideology, emotions, and values on the one hand, and, on the other, stark and naked matters of fact."[2] Christina Bieber Lake counters this separation with an insistence that instead of maintaining the stark division between fact and affect—or, as the debate has recently been framed, the sciences and the humanities—these two disciplines need to come into conversation with each other in order to ensure a balanced approach to future technologies. The limited engagement that literary studies now maintains with science and technology prevents a very real and important aspect of the human experience from coming into lasting and meaningful contact with the technological experience. Bieber Lake insists that this was not always the case; a return to serious engagement between literature and science would, indeed, be a return to the classical understanding between philosophy and society:

> For [ancient readers like Seneca and Augustine], the hard work began after the text was read, in applying its vision to gain self-awareness and to grow in the virtues. From that position individuals could make meaningful contributions to the *polis*. The modern reader, on the contrary, locates the work to "some form of interpretation, that is to expounding, clarifying, or explaining the text." The ancient question of the good life is left behind, and "interpretation has become the only widespread postreading activity."[3]

The result of this deferral to interpretation rather than an insistence upon a direct engagement with technology-through-literature via the question of ethics has resulted in important philosophical questions being delegated to scientists and politicians, who are not trained to interpret narrative to arrive at the answers to ethical questions.[4] For Bieber Lake, narrative is the key to an ethical approach to technology, to not only taking care of our "children" as they make their way in the world in their various and ever-changing forms, but ensuring that we are suitable parents. "Narrative does not visit ethical questions abstractly; it lives them, because it lives in the realm of ethos, of persons as persons engaged with one another."[5]

Through narrative we are able to live out the scenarios into which we will be (or are already in the process of being) thrust into by technology. Through narrative, we experience the emotional lives made possible by technology and live both the problems and solutions brought into being by that technology. Science fiction narratives especially, by launching the reader into futures and technologies that may as of now exist only as tiny ink blots, contemplative pauses in the scientist's notebook, demand an engagement across the spheres of fact and feeling. This chapter will demonstrate the link that science fiction creates between present and future, specifically in the realm of biotechnology and food security. Bringing the environmentalism of Vandana Shiva—and its awareness of the effects of Western imperialism, colonialism, and capitalism—to the Marxist implications of the death of the commons in the increasingly interior spaces of genetic material, this chapter examines current struggles in environmentalism and anticapitalism in the work of Paolo Bacigalupi, whose science fiction speaks to future manifestations of our present imbalances of power. In this way, this discussion establishes science fiction as a genre that demands a critical engagement with both lived experience of a future society and the present experience that points to it. Through an engagement that thoroughly considers the foundations that we, as the mothers and fathers of the technologies of the future, are establishing, we may be able to ascertain with greater accuracy whether that technology will itself be the parent of a utopian or dystopian world.

Before examining the work that science fiction does to map the trajectories of the present, it is first necessary to establish the best-case scenario for the future: if we heed literary warnings about our current course into the future, what futures are possible? What utopias fuel the hope of the science fiction writer? Utopian fiction as a science fiction subgenre was prevalent in the 1970s, with ecological and feminist futures envisioned among writers such as Marge Piercy and Ursula K. Le Guin. Raymond Williams groups these science fiction utopias into four distinct groups:

> (a) the paradise, in which a happier life is described as simply existing elsewhere; (b) the externally altered world, in which a new kind of life has been made possible by an unlooked-for natural event; (c) the willed transformation, in which a new kind of life has been achieved by human effort; (d) the technological transformation, in which a new kind of life has been made possible by a technical discovery.[6]

While the optimism of the 1970s has largely dissipated, Williams' categories remain relevant in their ability to isolate distinct elements of utopia that contemporary science fiction writers indirectly address. For the purposes of this chapter, we can discount Williams's second example, which is arrived at through serendipity; this utopia does not arise from anything that human beings have explicitly done themselves. The third and fourth utopias fit nicely together; indeed, Latour insists that this combination is likely the only possible salvation for a world in crisis. What are we to make of that classic model of utopia, the paradise that in many science fiction novels transports humanity away from otherwise persistent problems? I argue that all other utopias—and especially human-made utopias—are not possible without the mythological foundation of the paradise, a utopia that expresses the base elements of successful sociality: equality between the sexes, equilibrium in environment, and overall relational harmony. It is this utopia upon which all others are built. The paradise injects hope—that vital ingredient—into the social projects that work toward practical, human-made (and therefore flawed) utopia. With this utopia (and its working versions) in mind, we proceed to discuss the ways in which narrative connects the human creations of technology to the futures that they make possible.

There are currently many who are doing the work to bridge the gap between art and science in order to divert a dystopian trajectory, linking powerful symbols with their social manifestations and, in this way, provoking conversation with emergent technology. Vandana Shiva, the mother of an environmental movement that engages directly with the foundations of the society into which agricultural-biotech monoliths are releasing the products of their technology, connects imperialist and neoliberal politics with current biotechnologies. Shiva draws a parallel between the Columbus expedition and its resulting centuries of colonialism-via-religion to a more insidious colonization project that is currently ongoing through patents and intellectual property rights.[7] Just as Columbus and the Portuguese, French, and British who came after him viewed the Americas as an empty land that had not been claimed by the people who lived there, biotechnology firms are finding ostensibly unclaimed space in increasingly interior locations—inside the cells of plants, inside the human body, inside the DNA that makes up all life-forms. Shiva asserts that

> the takeover of native resources during colonization was justified on the ground that indigenous people did not "improve" their land. [...] The same logic is now used to appropriate biodiversity from the original owners and innovators by defining their seeds, medicinal plants, and medical knowledge as nature, as non-science, and treating the tools of genetic engineering as the yardstick of "improvement."[8]

Thus, the colonialist logic, that if a space is left in its natural state it is unclaimed property waiting to be possessed and plundered, is being applied to natural life-forms, which are then modified and claimed as property and through that process removed from the commons of biological diversity. Shiva goes on to say that

> through patents and genetic engineering, new colonies are being carved out. The land, the forests, the rivers, the oceans, and the atmosphere have all been colonized, eroded, and polluted. Capital now has to look

for new colonies to invade and exploit for its further accumulation. These new colonies are [...] the interior spaces of the bodies of women, plants, and animals.[9]

These colonized interior spaces become sites of violence, much like the violence that marks the colonial project in the Americas. "Patenting living organisms encourages two forms of violence. First, life-forms are treated as if they are mere machines, thus denying their self-organizing capacity. Second, by allowing the patenting of future generations of plants and animals, the self-reproducing capacity of living organisms is denied."[10] When intellectual property rights are applied to the genetic material that makes up plant, animal, or human life, that life becomes analogous to a factory that produces a specific product—a gene that has been modified for a singular purpose. "Some biologists have gone far into exalting the gene over the organism and demoting the organism itself to a mere machine. The sole purpose of this machine is its own survival and reproduction, or perhaps more accurately put, the survival and reproduction of the DNA that is said both to program and to 'dictate' its operation."[11] Shiva identifies reproduction—and the bodies of women—as a site of colonial violence. It is not *any* life that the new colonialism requires—it is fecund, reproductive life, life that will increase capital and thus power. For corporations whose product is rooted in the genetic manipulation of life, the ability to not only reproduce their product but also to have a monopoly on the reproduction of the product is essential to the expansion of empire.

Largely due to the work of decolonization and anti-GMO activists, biotechnology firms have had to slow their expansion into the internal realms of the modified genome and work around the sensitive issue of controlled reproduction. In 1999, biotechnology behemoth Monsanto announced that it would not commercialize their "terminator seed" technology[12]—a technology that prevented farmers from saving seed from one season to plant the next, thus chaining farmers to the corporation by the links of modified DNA that controlled reproduction and removed genetic material from the commons. Wilhelm Peekhaus describes this mechanism, through which capitalism becomes self-perpetuating. He states that "enclosing the commons augments the disciplinary processes of capital because such practices render greater numbers of people dependent upon the market in order to reproduce their livelihoods."[13] In light of this function, the announcement that Monsanto had abandoned its terminator seed technology was applauded by biotechnology critics who had projected the idea of the terminator seed into the future and envisioned a world in which formerly free technologies—the biological mechanism by which nutrients from the earth are transferred to human beings through seeds that are planted, germinated, grown, harvested, and eaten—are appropriated by corporate interests and thus are suddenly out of reach for a large section of the world's population.

The "terminator seed" technology, however, is alive and well in current agricultural biotechnology research. For example,

> Novartis, a Swiss biotech firm and the world's top seller of agrochemicals, was recently awarded a patent for a gene technology that would tie a whole set of plant development processes, including germination, flowering, and fruit ripening, to externally applied chemicals—perhaps even to Novartis's own proprietary chemicals.[14]

In this process, control over genetic material is ensured not through the genome itself, but in reaction to a chemical that one corporation holds like a key. Gabriela Pechlaner states that

> while agricultural biotechnologies deepen and extend the potential for appropriationism and substitutionism in agriculture, there are indications that this is not the only means by which they are facilitating capital accumulation. Jack Kloppenburg argues that capital accumulation in agriculture has been impeded by nature's obstruction to the commodification of the seed (that is, its reproducibility). As a consequence, capital has attempted to commodify the seed through two routes: one technical (i.e., physical impediments to reproduction, such as hybrid technology) and the other social (i.e., legislation to protect plant breeders, such as the Plant Variety Protection Act).[15]

Both technical and social routes to commodification are a concern not only for Western farmers, but also for developing countries where—if the reproductive aspect of plant life is removed from their control—agricultural production "could become wholly dependent upon foreign exports of critical chemical inducers. [...] Thus, genetic trait control technology could become a biological weapon used for agroterrorism."[16]

It is essential then that, in the face of these threats to food security and human welfare, this colonization of life itself is resisted. Vandana Shiva insists that a resistance against patenting of genetic material is a resistance of imperialism, a resistance that stands between the commodification of life and "the future of non-Western traditions of relating to and knowing nature. It is a struggle to protect the freedom of diverse species to evolve. It is a struggle to protect the freedom of diverse cultures to evolve. It is a struggle to conserve both cultural and biological diversity."[17] Any successful resistance movement draws from the iconography of the struggle and uses certain fecund images to reproduce and cultivate ideologies necessary to maintain the momentum of the movement; in the resistance of genetic imperialism, the only appropriate symbol is the seed. Shiva writes that

> the seed has become the site and symbol of freedom in the age of manipulation and monopoly of its diversity. It plays the role of Gandhi's spinning wheel in this period of recolonization through free trade. The *charkha* (spinning wheel) became an important symbol of freedom not because it was big and powerful, but because it was small; it could come alive as a sign of resistance and creativity in the smallest of huts and poorest of families. In smallness lay its power.
>
> The seed, too, is small. It embodies diversity and the freedom to stay alive. And seed is still the common property of small farmers in India. In the seed, cultural diversity converges with biological diversity. Ecological issues combine with social justice, peace, and democracy.[18]

The seed is a symbol of freedom because it has for so long—since the dawn of life on this planet—been free. It is a symbol of something that is available to all, not shut away in laboratories and accessible only to those with the means to pay for access. The seed is a symbol that echoes throughout all life, whether it be plant or animal, like a fractal that reaches each mimetic tendril into every facet of life as we know it, and into a future that becomes difficult to imagine without the seed. As such, it is a potent symbol that recurs in life and literature and allows for speculation that projects the technologies of the past—the simple agricultural implements that have fed humanity for thousands of years—into the possibilities of the future. The technologies that are being developed now in laboratories across the world *will* be born into the world. What can literature—the medium that is so often a conversation between symbol and materiality—say to its reception? How can science fiction bridge the gap between the seed and its emergent life?

When the question is one of the effects of biotechnology on the future of life, the narratives of speculative fiction can provide a kind of answer. These are the grounds on which science and imagination meet to discuss ethics, the landscape over which characters struggle in very real ways with the world they have been given. Nicola Griffith makes a comparison between science fiction and science itself: "Both are narrative," she stresses. "They are stories we tell to make sense of the world. Whether we're talking equation or plot, the story is orderly and elegant and leads to a definite conclusion."[19] Griffith asserts that science fiction engages readers not only with future worlds, but also with future selves. Science fiction often extends current social structures into frameworks of future possibility. Importantly, however, Griffith states that on a fundamental level, the best science fiction is about economies of love: love that enables conversation between the reader, the author, and a shared future—and a willingness to be changed by that experience. "Science fiction changes more than the world, more than our place in the world, it changes us [...], change[s] the discourse on what it means to be human. Through tall tales of human cloning, prosthetics, genetic engineering, it introduce[s] us to the notion that the nature of body and mind are mutable."[20]

Nancy Kress duly notes, however, that science fiction is not only a genre of possibility, but also a genre of risk. As a genre that relies mostly on plot-driven narratives, there is a negative bias in science fiction against scenarios that do not drive the narrative forward; in other words, science fiction readers do not want to read stories about successful technologies with only positive impacts on humanity.[21] It could be argued, however, that despite a focus on the negative impact of some technologies on the future world, positive technologies are seamlessly integrated into extra-narrative story elements, such as setting. These positive technologies, then, become a part of the considerations of both reader and character, as both are immersed in positive technology in a way that depends on the mutability of the human body and mind in much the same way as negative technologies, as a reality of that world.

Narrative, however, demands a problem to be solved; in science fiction, that problem is often one of a conflict between technology and humanity. In the biopunk subgenre of science fiction, these elements combine with the additional force of the natural world. Here we return to the seed as a symbol. The short stories and longer fiction of Paolo Bacigalupi take current ideas—in this case, Vandana Shiva's concern for intellectual property rights, colonialism, and resistance—and project them into the future. Bacigalupi begins with a seed—a small thing that nonetheless embodies the immense ideas of diversity and freedom—and through this symbol explores the details in relationships as intricate as the exchange of a single unit of energy. In many of the offerings from Bacigalupi's short story collection *Pump Six and Other Stories* and in his novel *The Windup Girl*, a colonized interior world presents the reader with a worst-case-scenario look at current biotechnology practices and the politics that surround them, and demands that we, the parents of this future world, answer for the transgressions of our technological children. Through science fiction, Bacigalupi holds up a mirror through which we can see not the negative consequences of advancing technology, but the alarming effects that current social structures have on increasingly powerful and complex science.

The world that Bacigalupi created for *The Windup Girl* first appeared in *Pump Six* in two short stories titled "The Calorie Man" and "Yellow Card Man." All three works present a world in which easy and cheap carbon-based energy sources are long gone and energy is measured in calories and joules—energy units consumed to support human life and energy units expended to maintain the machines that support the functions of a greatly diminished society. While "Yellow Card Man" and *The Windup Girl* detail a post-oil life in and around Bangkok, Thailand, "The Calorie Man" takes place in America, heading south along the Mississippi River, where fields of genetically modified SoyPRO and HiGro stretch verdant fingers tantalizingly to the horizon. It is a setting not unlike what one might encounter in many rural American spaces today. In fact, at first the only indication that this landscape is colonized space is the propaganda posters that the main character, Lalji, encounters: "One said: 'Unstamped calories mean starving families. We check royalty receipts and IP stamps.' [...] The other poster was AgriGen's trademarked collage of kink-springs, green rows of SoyPRO under sunlight and smiling children along with the words 'We Provide Energy for the World.'"[22] The posters identify the control that agricultural megacorporations exact over the land, power moving from their jurisdiction on the surface to the resources hidden in the interior spaces.

In "The Calorie Man," Bacigalupi focuses on the very element that Vandana Shiva finds so troubling; IP (intellectual property) agents patrol the areas surrounding the fields of SoyPRO and HiGro, collecting royalties on patented and highly regulated genetic material, thus colonizing not only the land but also the foundations of life. Lalji and his partner Creo take a job transporting a "generipper" (an extrapolation of the digital pirates that are ubiquitous today and universally loathed by media monoliths) downriver to safety so that he might restore fertility to the tightly controlled corporate fields. While the generipper, Bowman, represents a resistance movement that stands against the utter domination of an imperialist agricorp monopoly, "The Calorie Man" points to a colonization of ethics and morals—a colonization of the human soul—that is significantly more insidious. While biotech corporations clamp down on the creative force through the patenting of genetic material, the resulting loss of control on the ground level—where Lalji, Creo, and the generipper Bowman are precariously balanced between starvation and subsistence—produces a parallel effect of closing off to others just as the corporations close off the commons of the seed. When Lalji and his partner Shriram discuss taking Bowman onto their boat, the closed nature of relations in this world of colonized external and internal spaces becomes apparent. Lalji simultaneously recognizes the need for resistance and succumbs to an insular protectionism. Lalji questions, "'And just because he is an enemy of AgriGen I should help him? Some former associate of the Des Moines clique? Some ex-calorie man with blood on his hands and you think he will help you make money?' Shriram shook his head. 'You make it sound as if this man is unclean.'"[23] The use of the word "unclean" in the context of a society that is outwardly multicultural but also thoroughly colonized hints at a tentative acknowledgement of the spiritual realm and seed of the human spirit. Lalji's hesitancy points to the effect that such environmental manipulations have on the soul and its extensions of compassion, love, and understanding. Lalji cannot empathize with Bowman because the powers of the world he lives in have a monopoly on the creative force. The colonization of the seed—and by extension the source of life—by biotech firms results in a stunted expression of humanity and, thus, inadequate human reactions. The soul, like the seed, cannot resist colonial powers. The patenting of genetic material has a cascading effect, resulting in an insular, individualistic society that can no longer rely on the creative commons.

While "The Calorie Man" extends the idea of genetic patents and their effects on humanity to one small group of people, Bacigalupi's *The Windup Girl* develops this idea further by bringing it into an arena of intricate international politics. Again, Bacigalupi projects current political movements—the increased influence of Monsanto and other agrochemical corporations on the developing world—onto the landscape of a future world. A 2009 article in *The Economist* cites that while "America has 250,000–300,000 active farmers[,] India has 15 [million] cotton farmers alone, several million of whom Monsanto says it has reached already" spreading not only their GM seed, but also a "new Violator Exclusion Policy [that] denies farmers who break the terms of its licences access to all its technology for ever."[24] It is not in the breadbasket of rural America, then, that biotechnology will make (and is already making) its greatest inroads, but in developing countries like India or, as in *The Windup Girl*, Thailand.

Likewise, it is not in the developed world that colonial powers can find resources ripe for exploitation, but in the developing world to which megacorporations have not yet extended their power. Indeed, the premise of *The Windup Girl*—that Anderson, a "calorie man" (a spy and scout for a big agricultural corporation), is stationed in Bangkok undercover as a kink-spring (or electrical generation facility) owner while secretly searching for the untampered genetic material of Thailand's National Seed Bank—suggests that not only is the developing world essential for the new colonial project, but that this project is undertaken entirely covertly while maintaining a structure in support of old powers. When Anderson, who scours the Bangkok markets in search of "new" fruits and the seeds they contain, discovers the *ngaw* (rambutan) which was previously thought to be extinct, he is quick to claim his discovery. Bacigalupi writes,

> [Anderson] hands over a hemp sack without bothering to bargain. Whatever she asks, it will be too little. Miracles are worth the world. A unique gene that resists a calorie plague or utilizes nitrogen more efficiently sends profits sky-rocketing. If he looks around the market right now, that truth is everywhere displayed. The alley bustles with Thais purchasing everything from generipped versions of U-Tex rice to vermilion-variant poultry. But all of those things are old advances, based on previous genehack work done by AgriGen and PurCal and Total Nutrient Holdings. The fruits of old science, manufactured in the bowels of the Midwest Compact's research labs.[25]

Anderson's search for the National Seed Bank exemplifies corporate colonialism's insatiable drive to possess, to modify, and to claim for its own the internal spaces where life itself resides. Anderson turns his nose up at "old advances," the already claimed territory that is no longer under contest. Biotech colonialism is hungry for fresh genetic material; and when Anderson tastes the *ngaw* he finds the fruit to be "a fist of flavor, ripe with sugar and fecundity." He staggers in "the shell-shocked moment of flavor—real flavor—after a lifetime devoid of it."[26]

Indeed, it seems that the wild, uncontrolled genetics offered in the *ngaw* provide an experience that more closely maps to Anderson's remembered childhood experience of food, tapping into a nostalgic past that seems more real than the lived present. Even while recognizing the power of this "old science," however, Anderson pursues the enclosure of this genetic commons. He buys the sack of *ngaw* from the market not to enjoy what it has to offer, but to more easily track down the source of its genetic material, to move a step closer to uncovering the untapped resource of the seed bank. For Anderson and the other "calorie men" the value of the *ngaw* and, by extension, the seed bank, is not the association to the unfettered sensations and affect of childhood—or even a distant, mythical past wherein human freedom is expressed through interactions with a likewise autonomous environment—but through the power a select few can gain by locking down genetics with both technology and a new colonialist policy. Bacigalupi's narrative, then, serves as a warning against not only current corporate practice, but also the ways in which these practices have so thoroughly infiltrated and been accepted by Western culture.

The Windup Girl's explicit condemnation of agricultural biotech within the context of neocolonialism extends into the speculative with a commentary on similar technologies applied to human DNA. While the struggle between Thai nationals and the Western colonial project is expressed in *The Windup Girl* through control over the genetic possibilities contained in the Seed Bank, the long knives of colonialism have already sliced into the human genome to produce the New People, a race of beings used for tasks ranging from household servants and sex slaves to military-grade killing machines. Despite the fact that the title of the book suggests a mechanical nature for the new cyborg race, the New People are not mechanical, but a techno-biological manipulation of human and animal genes to create a physically superior people that will simultaneously be utterly but willingly enslaved by the physically inferior "old science" human beings.

Emiko, the only New Person directly in the novel, provides the reader not only with a view to the disturbing applicability of current technology in future worlds, but also signifies a commentary on current immigration policy and the devaluation and dehumanization of the other through capitalist commodification. Emiko is a household model who is abandoned by her Japanese master in Thailand when he discovers that it will be too expensive for him to bring her back to Japan upon his return. Emiko is no Harawayian cyborg; she struggles with both her identity and the duality of her nature. She simultaneously recognizes herself as a person deserving of rights and freedoms and also feels compelled to subjugate herself before others. She is at once disgusted by the biological functions of her genetically manipulated body and knows that she is biologically superior to all beings around her.

Unlike Haraway's cyborg, which "is not made of mud and cannot dream of returning to dust,"[27] Emiko *does* dream of Eden—an isolated place where New People can live free; though she doubts that the manifestation of this dream will ultimately satisfy her. Emiko's Eden—the mountains in which there is rumored to exist a colony of New People—provides an example of a society free of stratification, the single element that keeps Emiko and many others in Bacigalupi's world perpetually indebted and enslaved. Perhaps Emiko's conflicted nature serves to demonstrate, paradoxically, how remarkably human Emiko actually is; she is a more perfect expression of human nature than the humans that surround her, and there exists a hope that New People are capable of creating a world less prone to the corruptions of the old one.

Emiko's contradictions are also an indication of the violent war going on both around and inside her; just as Shiva points to the bodies of women as sites of colonial violence, Emiko's very body is a battleground in the colonial war. In conversation with Anderson about the technology of her genesis, Emiko recognizes that the colonial project has been built into every cell of her body. Bacigalupi writes,

> "Mmm, yes." Emiko's brow wrinkles thoughtfully. "[The genetically-modified cheshire cats] are too much improved for this world, I think. A natural bird has so little chance, now." She smiles slightly. "Just think if they had made New People first."
> Is it mischief in her eyes? Or melancholy?
> "What do you think would have happened?" Anderson asks.
> Emiko doesn't meet his gaze, looks out instead at the circling cats amongst the diners. "Generippers learned too much from cheshires."
> She doesn't say anything else, but Anderson can guess what's in her mind. If her kind had come first, before the generippers knew better, she would not have been made sterile. She would not have the signature tick-tock motions that make her so physically obvious. She might have even been designed as well as the military windups now operating in Vietnam—deadly and fearless. Without the lesson of the cheshires, Emiko might have had the opportunity to supplant the human species entirely with her own improved version. Instead, she is a genetic dead end. Doomed to a single life cycle, just like SoyPRO and TotalNutrient Wheat.[28]

Just as Bacigalupi's "calorie companies" manipulate the food supply through the genetic control of the reproductive qualities of seed, Emiko's genetics ensure that she and others of her kind provide a source of labor that will never supplant its human origins. Control over life is maintained through the curtailing of the reproductive impulse; even if Emiko were to find the secret colonies with which Anderson beguiles her, that freedom is a dead end without both the ability to reproduce in kind and nurture life into a quintessentially new way of thinking and being; without this capability, Emiko's people can have no lasting effect on the world. With total control of the production of New People, the expression of their personhood cannot reach its apex through a genome expressing its qualities in the wild.

John Harris suggests that a new breed of person is defined most clearly through reproductive capability. Harris states that status as a "breed" in and of itself "rests clearly but decisively on the fact that they could pass on this new constitution by normal reproduction with other members of the same 'breed.'"[29] Emiko's sterility, therefore, is a mark of the colonial project and its control over not only labor but also over the internal landscapes of both human and animal bodies. The fact that Emiko is made to submit to sex as part of her indentured labor that ensures her survival in Thailand is an expression of the colonialist hold over both internal and external bodily landscapes. That Emiko is used so brutally and so utterly without compassion for her personhood demonstrates the ways in which colonialism strips life down to its barest meaning and leaves nothing but resources to be consumed, spaces to be filled.

We need not, however, wait for technology advanced enough to bring New People into the world in order to see these effects already at work in the population. Control over reproduction of not only life but also of values and ideas meets little resistance in some segments of the population today. The combination of structures and circumstances are already in place, with some experiencing something very much like Emiko's degradation already in their own lives. These are themes, problems, and structures that Bacigalupi weaves into narrative repeatedly in both *The Windup Girl* and *Pump Six*. Even in stories that do not take place in the colonized space of the Post-Expansion world—such as in "The People of Sand and Slag" and "The Fluted Girl"—Bacigalupi revisits the ways in which biotechnological manipulations of the body's internal spaces amount to a loss of freedom that inhibits expressions of humanity, adeptly demonstrating how our present-day social structures have already ensured that these horrors will happen; we are simply waiting for the technology to catch up with us. For Bacigalupi, then, there is a danger that biotechnology could do the very opposite to what is most often posited; that is, the insidious and unchecked combination of biotechnology with neocolonial-corporate interest enslaves the bodies of plants, animals, and people, all while convincing us that what we drink so thirstily is progress and, thus, freedom.

While Bacigalupi's fiction does indeed present the negative bias that Kress warns about, it is not necessarily against the advances of biotechnology. Biotechnology has given Bacigalupi's world the cures to several plagues and blights and has created the New People, exemplified in Emiko, a physically superior being that demonstrates sensitivity and emotional depth that, in line with a classic cyborg trope, often outstrips that of the human beings around her. While Bacigalupi's treatment of biotechnology is certainly meant to inspire caution, that hesitancy is rooted not merely in the technology itself but also in the social systems into which the technologies are released. Read closely, it is not the future of biotechnology that provokes a fear response, but the future of that technology grafted onto neoliberal structures. One of the questions posed in both *Pump Six* and *The Windup Girl* is not only the effect of technology on our current social foundations, but also how such a grafting can be avoided. What ethic can help us to prepare our social infrastructures for the future that technology has to offer?

Here it is appropriate to return to the question of engagement with both the human experience of the emotional life and the rational adherence to fact demanded by work in scientific and technological fields. The balance of both these aspects of modern life necessarily requires an equally active conversation with the ethical questions requisite to meaningful involvement in both spheres. One cannot ethically engage in the work of bringing technological children into the world without also engaging in educational projects. Any such project should engage with both historical and present structures into which technologies will be inserted. Can technology lead us into a utopian society if it is merely grafted onto the old foundations that have produced wealth and prosperity for some, but slavery, starvation, and death for others? Perhaps in responding, it is wisest to begin by requiring an answer to the question "How should one live?" The use of the ambiguous pronoun here leaves open the position of subjecthood; in this opening one can see past an ethic of autonomy—a concern for a personal, individualistic freedom—to something more inclusive.

Bieber Lake enters into this conversation with Emerson and his ethic of individual freedom. She states that

> it is as if Emerson is saying, "what matters most is that I go forward with my freedom to change the world by starting with myself. Perhaps others have real claims on me, perhaps not, but let us go ahead and assume them to be real, and treat them well." But in a world where biotechnology is giving us startling new powers to literally and permanently shape the next generation, what does it mean to treat people well now? If I value my personal freedom highest, what happens to my neighbor? Can we simply

lay aside, as Emerson does here, the question of how real the other is? If we do, what in this philosophy will teach me what it means to treat someone well—to want for them the good life?[30]

Bieber Lake asserts that it is not autonomy that should act as the ethical rule by which to measure the technologies that we release into the world. The ethic of autonomy, unless purposefully extended to include the other, is myopically focused on the subject. It is this ethic, fixated on the power and affluence of Western economies, that drives the colonial project and any new colonialism to come. The higher ethic of love, on the other hand, engages both the I and the you, making subjects of both. Citing Habermas, Bieber Lake stresses that love as the guiding force of ethics has the ability to direct care and consideration toward the other. She stresses that "it is not genetic engineering that is the issue, but the attitude with which the interventions are carried out. In gene therapy, for instance, we have the correct attitude because we are thinking about the person the embryo will be. How we see the other person is the main thing that determines how we ascertain our duty to them."[31]

Perhaps most importantly, the ability of the love ethic to consider the other allows for the considerations of other-selves and other-futures that we simply cannot know with any precision at this time. Derek Parfit, in "Personal Identity," explored the idea that a person is comprised of multiple, separate selves and thus future selves take up the position of object rather than subject. This is a position that was confirmed in 2008 by Pronin et al. when experimental subjects committed their future selves to more arduous tasks in the name of science than they would commit to in the immediate present. This interesting quirk of human psychology suggests that it may be especially hard to consider future others in the projects we commit to today, as these are twice removed from our present selves. Adding to the difficulty are other factors such as the removals of geography, culture, religion, and sex that prevent an autonomy ethic from being an effective rule by which to measure the future of an increasingly globalized human experience. How, then, can love address the dangers that we may or may not face as we stand on the precipice of our biopolitical future? If, as Latour suggests, we must love our monsters and care for the technology we release into the world, we must also care for those whom our technology affects. This care demands that we exert ourselves today in examining and altering damaging social structures in preparation for our future children, technological or otherwise.

Notes

1. Bruno Latour, "Love Your Monsters," in *Love Your Monsters: Postenvironmentalism and the Anthropocene*, edited by Ted Nordhaus and Michael Shellenberger (Oakland: Breakthrough Institute, 2011), EPUB e-book.
2. Ibid.
3. Christina Bieber Lake, *Prophets of the Posthuman: American Fiction, Biotechnology, and the Ethics of Personhood* (Notre Dame: University of Notre Dame Press, 2013), xv.
4. Ibid., xiv.
5. Ibid., xvii.
6. Raymond Williams, "Utopia and Science Fiction," *Science Fiction Studies* 5, no. 3 (November 1, 1978): 203.
7. Vandana Shiva, *Biopiracy: The Plunder of Nature and Knowledge* (Boston, MA: South End Press, 1997), 2.
8. Ibid., 4.
9. Ibid., 5.
10. Ibid., 23.
11. Ibid., 29.
12. Brian Halweil, "Monsanto Drops the Terminator," *Worldwatch Institute* 13, no. 1 (2000): 8.
13. Wilhelm Peekhaus, "Primitive Accumulation and Enclosure of the Commons: Genetically Engineered Seeds and Canadian Jurisprudence," *Science & Society* 75, no. 4 (October 1, 2011): 544–555.
14. Brian Halweil, "Monsanto Drops the Terminator," *Worldwatch Institute* 13, no. 1 (2000): 8.
15. Gabriela Pechlaner, "Biotech on the Farm: Mississippi Agriculture in an Age of Proprietary Biotechnologies," *Anthropologica* 52, no. 2 (January 1, 2010): 293.
16. Mansir Yusuf, "Ethical Issues in the Use of the Terminator Seed Technology," *African Journal of Biotechnology* 9, no. 52 (December 2010): 8904, doi:10.5897/AJB2010.000–3320
17. Vandana Shiva, *Biopiracy: The Plunder of Nature and Knowledge* (Boston, MA: South End Press, 1997), 5.
18. Ibid., 126.
19. Nicola Griffith, "Identity and SF: Story as Science and Fiction," in *SciFi in the Mind's Eye: Reading Science through Science Fiction*, edited by Margret Grebowicz (Chicago: Open Court, 2007), 139.
20. Ibid., 143.
21. Nancy Kress, "Ethics, Science, and Science Fiction," in *SciFi in the Mind's Eye: Reading Science through Science Fiction*, edited by Margret Grebowicz (Chicago: Open Court, 2007), 203.
22. Paolo Bacigalupi, "The Calorie Man," in *Pump Six and Other Stories* (San Francisco: Night Shade Books, 2008), EPUB e-book.
23. Ibid.

24. "The Parable of the Sower," *The Economist*, November 19, 2009, http://www.economist.com/node/14904184.
25. Paolo Bacigalupi, *The Windup Girl* (San Francisco: Night Shade Books, 2009), EPUB e-book, chap. 1.
26. Ibid.
27. Donna Jeanne Haraway, *Simians, Cyborgs, and Women: The Reinvention of Nature* (New York: Routledge, 1991), 151.
28. Paolo Bacigalupi, *The Windup Girl* (San Francisco: Night Shade Books, 2009), EPUB e-book, chap. 10.
29. John Harris, *Wonderwoman and Superman: The Ethics of Human Biotechnology* (Oxford: Oxford University Press, 1992), 185.
30. Christina Bieber Lake, *Prophets of the Posthuman: American Fiction, Biotechnology, and the Ethics of Personhood* (Notre Dame, IN: University of Notre Dame Press, 2013), 3–4.
31. Ibid., 6.

Bibliography

Bacigalupi, Paolo. *Pump Six and Other Stories*. San Francisco: Night Shade Books, 2008. EPUB e-book.

———. *The Windup Girl*. San Francisco: Night Shade Books, 2009. EPUB e-book.

Bieber Lake, Christina. *Prophets of the Posthuman: American Fiction, Biotechnology, and the Ethics of Personhood*. Notre Dame, IN: University of Notre Dame Press, 2013.

Emerson, Ralph Waldo. *Nature and Selected Essays*, edited by Larzer Ziff. New York: Penguin, 2003.

Griffith, Nicola. "Identity and SF: Story as Science and Fiction." In *SciFi in the Mind's Eye: Reading Science through Science Fiction*, edited by Margret Grebowicz, 139–143. Chicago: Open Court, 2007.

Habermas, Jürgen. *The Future of Human Nature*. Malden, MA: Blackwell, 2003.

Halweil, Brian. "Monsanto Drops the Terminator." *Worldwatch Institute* 13, no. 1 (2000): 8.

Haraway, Donna Jeanne. *Simians, Cyborgs, and Women: The Reinvention of Nature*. New York: Routledge, 1991.

Harris, John. *Wonderwoman and Superman: The Ethics of Human Biotechnology*. Oxford [England]; New York: Oxford University Press, 1992.

Kloppenburg, Jack Ralph. *First the Seed: The Political Economy of Plant Biotechnology, 1492–2000*. 2nd ed. Madison, WI: University of Wisconsin Press, 2004. http://hdl.handle.net/2027/heb.06255.

Kress, Nancy. "Ethics, Science, and Science Fiction." In *SciFi in the Mind's Eye: Reading Science through Science Fiction*, edited by Margret Grebowicz, 201–209. Chicago: Open Court, 2007.

Latour, Bruno. "Love Your Monsters." In *Love Your Monsters: Postenvironmentalism and the Anthropocene*, edited by Ted Nordhaus and Michael Shellenberger. Oakland: Breakthrough Institute, 2011. EPUB e-book.

Parfit, Derek. "Personal Identity." *The Philosophical Review* 80, no. 1 (January 1, 1971): 3–27. doi:10.2307/2184309.

Pechlaner, Gabriela. "Biotech on the Farm: Mississippi Agriculture in an Age of Proprietary Biotechnologies." *Anthropologica* 52, no. 2 (January 1, 2010): 291–304.

Peekhaus, Wilhelm. "Primitive Accumulation and Enclosure of the Commons: Genetically Engineered Seeds and Canadian Jurisprudence." *Science & Society* 75, no. 4 (October 1, 2011): 529–554.

Pronin, Emily, Christopher Y. Olivola, and Kathleen A. Kennedy. "Doing unto Future Selves as You Would Do unto Others: Psychological Distance and Decision Making." *Personality & Social Psychology Bulletin* 34, no. 2 (February 2008): 224–236. doi:10.1177/0146167207310023.

Schubert, Robert. "Farming's New Feudalism: Percy Schmeiser and Other Casualties of Industrial Agriculture's Drive to Own It All." *Worldwatch Institute* 18, no. 3 (2005): 10–15.

Shiva, Vandana. *Biopiracy: The Plunder of Nature and Knowledge*. Boston, MA: South End Press, 1997.

"The Parable of the Sower." *The Economist*, November 19, 2009. http://www.economist.com/node/14904184.

Williams, Raymond. "Utopia and Science Fiction." *Science Fiction Studies* 5, no. 3 (November 1, 1978): 203–14.

Yusuf, Mansir. "Ethical Issues in the Use of the Terminator Seed Technology." *African Journal of Biotechnology* 9, no. 52 (December 2010): 8901–8904. doi:10.5897/AJB2010.000–3320.

PART III

Reactions

CHAPTER 6

"All Day, All Week, Occupy Wall Street!": Space, Biopower, and Resistance

Elena L. Cohen

The defining feature of the so-called Occupy movements that emerged worldwide in 2011 is implicit in the name itself: these protests are a call to *occupy physical spaces*. Indeed, the original flyer for Occupy Wall Street bore only one instruction: "BRING TENT."[1] Following this call to action, an encampment was established in Zuccotti Park, in New York City's Financial District.[2] In both academic and popular writings, "Zuccotti Park" has been synonymous with Occupy Wall Street, and after its "eviction"[3] many question(ed) whether the movement even still exists.[4] However, before and after the raid on Zuccotti Park, many large Occupy Wall Street actions occurred in spaces apart from the park, and continue presently and into the foreseeable future.[5] The relationship between Occupy Wall Street and space is therefore a complex one, revolving around issues of knowledge, power, resistance, sovereignty, security, discipline, governmentality, and biopolitics more generally. Thus, Foucault's intricate views of these topics as they relate to space may be seen as an ideal lens through which to analyze Occupy Wall Street.

Foucault's discussions of space have been utilized previously in order to analyze other social movements, in particular the 2004 Republican National Convention protests and the 1999 Seattle World Trade Organization protests.[6] However, much of this work focuses exclusively on Foucault's discussions of the relationship between biopolitics and space, without raising the ways in which Foucault sees space as having the power to liberate and function as anti-hegemonic spaces where difference(s) can (co)exist.[7] This limited application of Foucault's views on space is perhaps unsurprising, as it reflects the similarly limited scope of works that discuss Foucault's views of space in more general senses.[8] Yet Foucault does argue, repeatedly and explicitly, that space, as both a tool and object of analysis, can be part of resistance and freedom,[9] and these views of Foucault can therefore be used to explore the liberatory possibilities of space in general, and Occupy Wall Street in particular. One particularly useful concept for understanding the positive abilities of space is that of "heterotopias," which Foucault defines as real places through which hegemony and domination can be resisted and differences can be affirmed.[10] Indeed, as will be argued below, Occupy Wall Street may be considered precisely one of these hegemony-challenging heterotopias of resistance.

To this end, the section "Space is Fundamental in Any Exercise of Power" will first briefly explain Foucault's views on the relationship between space and biopolitics, and will explain how the spatial dimensions of biopolitics have been used by those seeking to suppress protest movements and by protest movements themselves. The third section, "As a Sort of Simultaneously Mythic and Real Contestation of the Space in Which We Live," will lay out in detail Foucault's work on heterotopias and other ways in which space may be seen as liberatory. This section will then argue that Occupy Wall Street can best be seen as a space of resistance in general, and more specifically as a heterotopia. This chapter concludes that not only may Occupy Wall Street, and other protest movements like it, be considered non-hegemonic heterotopias, but may also, in accord with Foucault's views, be able to provide some sort of liberation or freedom.

"Space Is Fundamental in Any Exercise of Power":[11] Michel Foucault on Space and Biopower

Michel Foucault raises concepts of space, architecture, and place in many of his works, both as primary areas of concern[12] and in furtherance of other inquiries of study.[13] While much of his theorizing about space revolves around the exertion of power and discipline, important aspects of Foucault's work on space highlights its interplay with resistance and possibly liberation. As such, the next section highlights Foucault's discussion of space as it relates to biopolitics and the ways in which Foucault's ideas have been applied to political and social movements. The following section explores Foucault's work on the relationship between space, resistance, and liberation.

Space, Biopolitics and the Suppression of Occupy Wall Street

Foucault writes often and explicitly of the link between biopower and space, stating points that a "whole history remains to be written of spaces—which would at the same time be the history of powers"[14] and that discipline is "above all an analysis of space."[15] As such, space is an important element of Foucault's concept of biopolitics, in which individual bodies are subjugated and entire populations are controlled.[16]

Foucault's ideas of space as they relate to discipline and power have been widely used to analyze the response of the police and the legal system to social and political protests.[17] Much of this literature has sought to show the ways in which US First Amendment doctrine is flawed for considering space to be a neutral, static factor when regulating free speech, thus allowing most restrictions on space, as it relates to political speech, to be found constitutional.[18]

First Amendment jurisprudence, which defines the contours of how expressive activity may be regulated, focuses on *what* speech and actions are protected from government regulation, and *why* the government seeks such regulation.[19] *Where* expressive activity occurs is generally seen to be unimportant, with all spaces broken up into a binary of "public forum" versus "non-public forum," with speech (potentially) protected by the Constitution only in the former.[20] Even within traditional public fora, courts typically find that the government's (often poorly developed) interests outweigh the rights of an individual to express herself in a particular place, and/or that there are alternative places where the individual could express herself without any effect on the message.[21]

Instead of this neutered view of space, some legal scholars have urged that the courts take into account Foucault's view of space as a dynamic and active force that can control and discipline bodies and populations.[22] For example, attention has been drawn to the ways in which free (political) speech has been "disciplined, controlled, and even suppressed through a variety of special tactics."[23] As evidence, attention has been drawn to the confinement of protesters away from the events that they are protesting in the forms of "free speech cages," "restricted zones," and "frozen zones," and to the widespread use of metal barricades, nets, and laws regarding the use of the sidewalks to control the movement of protesters in general.[24] These tactics have been used against Occupy

Wall Street protesters, perhaps most infamously and recently with the example of the "1st Amendment Rights Area" / "freedom cage" in front of Federal Hall (see Figure 6.1).[25]

Indeed, the suppression of Occupy Wall Street illuminates many of the ways in which Foucault posits that space can be a tool of biopolitics. For example, in *Discipline and Punish*, Foucault discusses the relation between "power of place" and biopolitics, focusing on "spatial tactics."[26] The primary method of the modern state's exercise of control with place is its "spatial canalization of everyday life," that is its "binary division and branding,"[27] which allows the state to maintain order,[28] neutralize dissent,[29] and generally to create docile bodies.[30] As such, space functions for Foucault within a disciplinary society to provide "order, control, surveillance, separation, and branding."[31] This is achieved partially through the knowledge gained by the "categorization, classification, order, division, and hierarchy" of bodies, which generates knowledge that "establishes a regime of truth in which power is guaranteed."[32] Relating this to Occupy Wall Street, the New York Police Department (NYPD) and the Federal Bureau of Investigation (FBI) engaged in binary division and branding by classifying the Occupiers as "domestic terrorists" (as opposed to citizens exercising rights to protest), and was able to justify monitoring the Occupiers on the basis of this division and branding.[33]

Space also enters into Foucault's conceptualization of biopower in his lectures published as *Security, Territory, Population*, in which he argues that *sovereignty, discipline,* and *security* all share problems of space. [34] Turning first to sovereignty, this concept is perhaps most clearly linked with space as it is "exercised within a territory."[35] Sovereignty played a complex role in attempts to end the encampment of Zuccotti Park, with the NYPD unable to evict the protesters earlier, as the park was zoned as a "privately owned public space" that must remain open to the public for 24 hours a day.[36]

Turning to discipline, this concept is also spatial for Foucault, but moves from control over a space to the *structuring* of a space,[37] best seen within a framework of *construction*.[38] Discipline involves an *empty space*, or the purposeful *emptying of a space*, in order to ensure ideal functioning in four main areas: (1) hygiene, (2) trade within the town, (3) trade outside of the town, and (4) surveillance.[39] This discussion of Foucault's conceptualization of discipline as the purposeful emptying of a space is incredibly germane to the response to Occupy Wall Street, as the mayor of New York City, the NYPD, and the owners of Zuccotti Park repeatedly attempted to evict the protesters, and, once evicted,

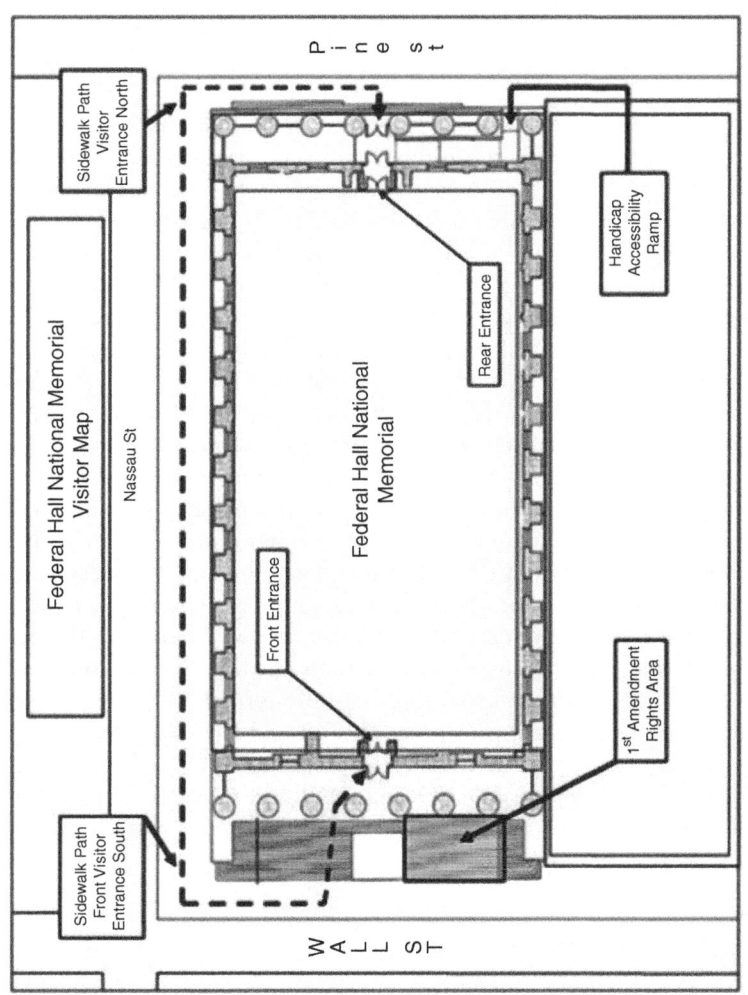

Figure 6.1 Notice of temporary access change.

put in place myriad regulations (backed by city officers and privately hired security guards) to ensure that the Occupiers could not return to the park. The four goals of discipline are all also present within the response to Occupy Wall Street: hygiene concerns were the purported reason why the park had to be cleared; business concerns both within and without New York City were clearly at issue; and, as discussed above, surveillance of Occupy Wall Street was a constant reality and concern.[40]

For Foucault, security also has important spatial aspects, and can be contrasted with the spatiality in four main ways: security (1) works on already existent spaces,[41] (2) is concerned with achieving goals similar to discipline, but without aiming at ideals,[42] (3) is concerned with polyfunctionality,[43] and (4) looks to the future, concerned with the temporal and uncertain, which Foucault refers to as the "milieu."[44] Security can be seen as having functioned as an apparatus within Zuccotti Park in these ways, as the response was to an already existent space (an Occupied Zuccotti park), with the same the goal of the disciplinary measures (ending the protest), but without complete idealism. Mayor Bloomberg knew that in clearing Zuccotti Park of protesters it would likely deal a heavy blow to the Occupy movement, but it is doubtful that he (or anyone else) thought that by controlling this space all protest and dissent in New York City would come to an end. Lastly, the response to Occupy Wall Street looked to *future* biological concerns in asserting its claims of security. Tellingly, at a press conference the morning after the eviction from the park, the then mayor, Bloomberg, stated that he ordered the park to be cleared because he had "become increasingly concerned—as had the park's owner, Brookfield Properties—that the occupation was coming to pose a health and fire safety hazard to the protesters and to the surrounding community."[45] Clearly, then, one way in which the suppression of Occupy Wall Street can be understood is through these spatial dimensions of biopolitics.

Occupy Wall Street's Use of Spatial Biopolitics

However, Occupy Wall Street's decision to set up an encampment of protesters can be seen as an attempt to use spatial tactics of their own: a form of self-branding that exploited the gaps of a First Amendment that finds space to be neutral, stymieing the efforts of law enforcement officials to neutralize the protest through their spatial tactics, and allowing the occupation of an area in an incredibly busy part of one of the busiest cities in the world to remain for almost two months.

If, as is argued above, space functions for Foucault within a disciplinary society to provide "order, control, surveillance, separation, and branding,"[46] then Occupy Wall Street, in (partially) separating, branding, and opening itself up can be seen as having used the spatial tactics of a disciplinary society, but for its own ends.[47] Certainly, the encampment of Zuccotti Park may be considered a type of self-separation, a call for people to physically leave their houses and instead remain within the park's confines, a place that would be filled exclusively with "Occupiers." Similarly, Occupy Wall Street can be seen as having branded itself, declaring the Occupiers to be representative of the "99%," as opposed to the "1%."[48]

In terms of surveillance, Occupy Wall Street publicized the meeting times of every working group both on boards in the park and on its website, and anyone was allowed to attend General Assemblies (see Figure 6.2). As Occupy Wall Street was not allowed to use amplified sound because they lacked the requisite city permits, they used a human microphone system, referred to as "the people's mic," in which one person would speak, and all others within earshot would repeat.[49] This

Figure 6.2 Zuccotti Park.

system ensured that important discussions of the group could be easily heard by all present—from fellow Occupiers and passersby to the massive number of police officers constantly circling Zuccotti Park. Combined with this massive militarized police presence around the park at all times, all involved were likely aware that all major decisions and actions of the group were being closely watched (see Figure 6.3).[50] There was, additionally, a large degree of control on those Occupying the park: meals were served at certain times and a quiet hour was enforced. In these ways, Occupy Wall Street varies from other large protests that have faced suppression. Nets, cages, and metal barricades were not able to neuter dissent and prohibit expressive activity when used against a movement that had (metaphorically and physically) set itself up within its own freedom cage.[51]

However, classifying Occupy Wall Street as self-branding, self-separating, self-revealing, and self-controlling is not the best way to understand either the movement or its relation to Foucault's concepts of space and power. While calling for the "99%" to Occupy Zuccotti Park, anyone was (theoretically) allowed into the park during its encampment. Similarly, although the "99%" versus "1%" concept is a type of

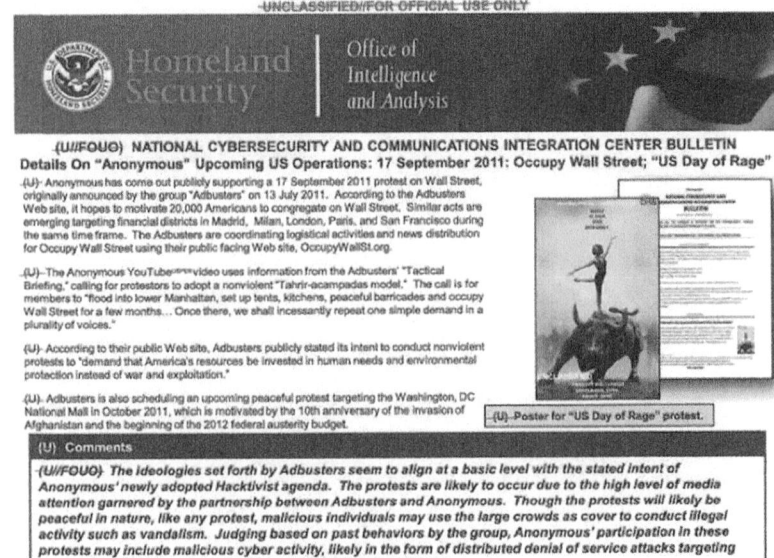

Figure 6.3 Homeland security intelligence document.

branding, there had been (and continues to be) a diversity of messages and modes of messaging within the movement.[52] Surveillance is a complicated notion as well, as it relates to Zuccotti Park, due to the presence of many tents, which made possible undetected and unobservable meeting spaces. Additionally, the in-person (as opposed to online) meetings perhaps made police surveillance more difficult, as an undercover officer would have to be physically present at each meeting.

In sum, the relations between Occupy Wall Street and the spatial tactics (usually) used to exert power over protesters are complex, and perhaps inverted in so far as the movement and in particular its encampment at Zuccotti Park actually did reappropriate tools meant to stifle dissent in order to allow dissent to flourish. What is clear from this discussion, though, is that much as the state has been charged with "actually, in some cases, creating places for the express purpose of controlling and disciplining protest and dissent,"[53] Occupy Wall Street created a space of its own for dissent and resistance. However, rather than examine it as a space of biopower (albeit used by protesters along with those seeking to suppress them), Occupy Wall Street can be better understood as a space of resistance, to which this chapter now turns.

"As a Sort of Simultaneously Mythic and Real Contestation of the Space in Which We Live":[54] Foucault on Space and Resistance

An Introduction to Heterotopias and Resistance

In addition to his musings on spatial tactics at the intersections of space and biopolitics, Foucault writes of space in relation to resistance and freedom.[55] Crucial to understanding Foucault's position on the liberatory possibilities of space is his elucidation of a concept that he terms "heterotopias," which are non-hegemonic real places existing within the network-like modern world and, as will be discussed below, are an ideal framework within which to view the Occupation of Zuccotti Park.[56] In *Of Other Spaces*, Foucault contends that the twentieth century may be considered an "epoch of space," concerned with "simultaneity," as a "network that connects points and intersects with its own skein."[57] Tracing the Western history of space, Foucault posits that the Middle Ages exhibited a "space of emplacement," which was hierarchical and about localizing things into their places.[58] This view of space was upset by the work of Galileo, who cast space as *infinitely open*, with "a thing's place...no longer anything but a point in its movement," thus creating a "space of extension" that would be prominent throughout the seventeenth century.[59] This in turn gave way to the twentieth century's understanding of space, as the "space of the site," with sites being "relations of proximity between points or elements," concerned primarily with *demographics*, that is the problem "of knowing...what type of storage, circulation, marking, and classification of human elements should be adopted in a given situation, in order to achieve a given end."[60]

Minor to Foucault's discussion on heterotopias, but quite relevant to the encampment at Zuccotti Park, is his observation that in the move from the space of emplacement to the space of the site via the space of extension, "contemporary space is perhaps still not entirely de-sanctified."[61] In other words, despite today's omnipresent networks of relations that challenge finiteness and circumscription, space in the twentieth century retains "a certain number of oppositions that remain inviolable, that our institutions and practices have not yet dared to break down," such as public versus private space and spaces of work versus spaces of leisure.[62] Instead of being challenged, these oppositional views of space are, Foucault argues, "nurtured by the hidden presence of the sacred."[63]

Turning to heterotopias themselves, Foucault posits that the spaces in which we live are neither empty nor homogeneous.[64] Instead, contemporary spaces are heterogeneous and reflect our age of space as sites, in so far as "we live inside a set of relations that delineates sites which are *irreducible to one another* and *absolutely not superimposable on one another.*"[65] These spaces may be seen as *heterotopias*, contrasted with utopias in that they are real places that actually exist, but are outside of all places, in that within them the "other real sites that can be found within the culture, are simultaneously *represented, contested*, and *inverted.*"[66] As will be discussed below, this composition of heterotopias by separate sites that remain distinct from each other while being in a complex set of relations with each other and with the heterotopia as a whole is crucial to understanding the dynamics of Occupy Wall Street.

In order to elucidate the most salient aspects of these heterotopias, Foucault suggests a "heterotopology" of six main principles. First of these is that heterotopias exist in every culture, but in different forms, and have shifted from being dominantly "crisis heterotopias" to the current domination of "heterotopias of deviation."[67] While the former reserves spaces for those seen as in a state of crisis within the culture (such as children or pregnant women), the latter is where "individuals whose behavior is deviant in relation to the required mean or norm are placed" and includes spaces like prisons and psychiatric hospitals.[68]

Second, each heterotopia has a "precise and definite function within a society," but this function can change and can be different for similar heterotopias within different cultural contexts.[69] Third, within a heterotopia there can be other real sites that are otherwise "incompatible" and "contradictory" with each other.[70]

Fourth, heterotopias are often linked to specific moments in time, or what Foucault refers to as "heterochronies,"[71] and they begin to operate fully only when "men [sic] arrive at a sort of absolute break with their traditional time."[72] There are two major ways in which heterotopias are related to time: (a) those that are concerned with the "perpetual and indefinite accumulation of time in an immobile place," which seek to archive " all epochs, all forms, all tastes, the idea of constituting a place of all times that is itself outside of time and inaccessible to its ravages";[73] with the prime examples of these being museums and libraries, Foucault sees these accumulative heterotopias as integrally related to Western modernity, and (b) those that are more ephemeral and transitory, in the "mode of the festival."[74] Crucially for the discussion of Occupy Wall Street below, Foucault notes that these spaces of festival and accumulation can come together, in some cases abolishing time as they rediscover it.[75]

Fifth, heterotopias are usually not freely open and accessible, but "always presuppose a system of opening and closing that both isolates them and makes them penetrable."[76] Entrance into a heterotopia is, according to Foucault, premised on the passage of some rite, of getting permission, of going through certain motions. Even heterotopias that appear to be fully open are illusions, hiding the secret ways in which, although one believes one has entered, one is actually excluded.[77] Lastly, heterotopias have one of two functions "in relation to all the space that remains": they are either "heterotopias of illusion," which reveal that all "real" sites where human life is partitioned are actually illusions, or "heterotopias of compensation," which strive to create idyllic states of planned perfection that are sharply contrasted with our more common disorderly and unkempt everyday sites of life.[78]

Foucault concludes this article with what comes across as plea for the establishment of many heterotopias across many societies and cultures.[79] Although throughout the work many of the heterotopias described by Foucault seem to be imbued with negativity (prisons, nursing homes, hospitals), at the end of this piece Foucault posits the ability of heterotopias to be much more positive entities, through which hegemony and domination can be resisted and differences can be affirmed.[80] Commentators to this work have therefore referred to these heterotopias as "lived utopia[s],"[81] and "place[s] where alternatives are considered, 'common sense' is questioned and business as usual stops for a moment."[82] Indeed, as will be argued below, Occupy Wall Street may be considered precisely one of these hegemony-challenging heterotopias of resistance.

The notion that a place can serve a liberatory function is raised by Foucault in other works as well.[83] Although Foucault did not believe that the architecture of a place could lead to a lasting freedom, he argues that structurings of space can have a liberatory effect, if combined with actual and constant practices of liberation within this constructed space. In response to a question posed by Paul Rabinow as to whether architecture can itself "resolve social problems," Foucault responds that "architecture can and does produce positive effects when the liberating intentions of the architect coincide with the real practice of people in the exercise of their freedom."[84] Liberty, for Foucault, is therefore a *practice*, and only exists in places where liberty is actively exercised.[85] Yet, in this view, freedom *can* exist in a space constructed for that purpose—so long as it is coupled with active efforts in the pursuit of this freedom.

Additionally, Foucault argues that even space exercised in a disciplinary manner can be co-opted by those over whom the power is exerted. "Spaces of captivity," he writes, "can be reversed: the inside outside."[86] That Foucault sees a space as able to liberate, at least in part, is crucial to understanding Occupy Wall Street's encampment of Zuccotti Park. Indeed, the liberatory function that Foucault offers in a limited way to space gives weight to the claim advanced below that not only did Occupy Wall Street form a sort of heterotopia of resistance, but that this occupation could, in reality, have a lasting and positive impact on freedom.

Occupy Wall Street as a Liberatory Heterotopia

Occupy Wall Street is a protest movement. Although it has been accused of not having "demands," one dominant characteristic is clear: all involved in the Occupy movements are in some way opposed to the current practices of the society around them.[87] As a space of resistance and deviation from society, Occupy Wall Street can be seen as a heterotopia, the non-hegemonic sites where difference can exist described above. Indeed, *representation*, *contestation*, and *inversion* of other real sites from the dominant culture offer an ideal way in which to view the activities inside of Zuccotti Park and the Occupy movement.[88]

Before moving through Foucault's heterotopology, it is interesting to note that Occupy Wall Street moved relatively far from a space of emplacement to a space of the site, through its explicit challenge of many "sacred" boundaries.[89] The "inviolable" division between public and private can be seen as being challenged, for example, by the encampment of Zuccotti Park, in which Occupiers slept, ate, and occasionally had sex—core "private" activities in our culture.[90] Similarly, the seemingly sacrosanct space between work and leisure is challenged through the Working Groups, ranging from Campaign Finance Reform group and Queer Caucus to the Kitchen, Medics, and Finance, in which participants are involved with unpaid, self-selected activities that they are often personally interested in, but often with large time commitments and in lieu of other "occupations."[91]

Turning to Foucault's principles of heterotopias, under the first principle, Occupy Wall Street may easily be seen as a heterotopia of deviation, which is characteristic of contemporary heterotopias, in so far as these heterotopias are ones in which "individuals whose behavior is deviant in relation to the required mean or norm are placed."[92] As discussed above, the core uniting factor of Occupy Wall Street is that, despite divergent specific goals and beliefs, all share a common resistance to the status quo.[93] It is of immense importance, however, that the Occupiers have *placed themselves* into this space of common deviance.

Further, under Foucault's second principle of heterotopias, Occupy Wall Street can be seen as having a "precise and definite function within a society," albeit one that is open to change, in much the ways anticipated by Foucault.[94] Indeed, Occupy Wall Street has the precise and definite function of calling into question the economic, social, and cultural factors that underlie a system in which 1 percent of the population controls a grossly disproportionate amount of the society's resources.[95] This function of questioning the status quo shows how, as Foucault explains, similar heterotopias can function in different ways in different places: the occupation of Wall Street arguably had different functions than the occupation of Tahrir square.

As for Foucault's third principle of heterotopias, Occupy Wall Street encompasses the precise sorts of "incompatible" and "contradictory" other sites that Foucault contemplates.[96] In this movement, "black bloc" anarchists occupy alongside those who advocate reform of the system along with those who would simply like this system to function fairly, along with those who are not sure exactly what they want yet hold the conviction that the status quo is not it.[97] In this way, much like the

Oriental gardens in which the four corners of the world, and their disparate symbols and meanings, were meant to converge, Occupy Wall Street can be seen as the "smallest parcel of the world and then it is the totality of the world."[98]

Continuing with Foucault's heterotopology, Occupy Wall Street works well with the fourth principle of heterotopias, which discusses temporality. As with other heterotopias, Occupy Wall Street began "to function at full capacity when [people] arrive[d] at a sort of absolute break with their traditional time."[99] Foucault posits that heterotopias can be either (1) perpetual, immobile places or (2) more temporal, festival like spaces, both of which were present in Occupy Wall Street's encampment of Zuccotti Park. Occupy Wall Street's immobile and perpetual aspects include its library, which sought to accumulate an indefinite number of works,[100] and its website that stores an ever-expanding network of information related to the movement.[101] Additionally, the movement has continued to thrive after its eviction, drawing tens of thousands to marches across the Brooklyn Bridge, a "Re-occupation" in December, and the May Day protests the following year.[102]

Turning to the latter mode, however, Occupy Wall Street also exhibits its fleeting and spectacular qualities. Indeed, "mode of the festival" would likely be a term that many would find appropriate to describe Occupy Wall Street, even if uninitiated with the Foucauldian sense.[103] This blending of modes is not a challenge to its classification as a heterotopia, as Foucault allows for this occurrence, himself referring to the example of "exotic" vacation villages that have been invented to "offer a compact three weeks of primitive and eternal nudity to the inhabitants of cities."[104] So too can Occupy Wall Street be seen as giving a crash course in activism, particularly when the encampment at Zuccotti Park was still in existence, in which people with all sorts of activist and non-activist backgrounds could literally eat and sleep within the movement.

The fifth principle, that of non-free entry, is perhaps the most obtuse with regards to Occupy Wall Street, as it is not clear that the movement requires a type of permission or rite of passage to enter.[105] However, there are certain motions that must be practiced in order to participate with the movement, among them the literal hand motions used to vote in general assemblies.[106] The final principle of heterotopias has a much cleaner fit with Occupy Wall Street, as the movement may be seen as both a heterotopia of illusion, in that many of its actions seek to reveal that all "real" sites where human life is partitioned are actually illusions,[107] and as a heterotopia of compensation, that strives to create a space that is planned to be ideal in relation to our broken status quo.[108]

Therefore, not only does Occupy Wall Street fit well in Foucault's heterotopology, but the examination of the movement from this perspective elucidates many of its major aspects. Thus, an analysis of Occupy Wall Street within the context of Foucault's discussion of heterotopias provides an ideal way within which to bring to light its anti-hegemonic and pro-difference character. Coupled with the discussion above of Foucault's view that, with continual practice, a space can provide freedom,[109] the continuous actions of resistance by Occupy Wall Street may be able to deliver just such a liberation.[110]

Conclusion: Of Gardens, Boats and Protest Movements

Foucault concludes his essay on heterotopias by turning to boats. He writes:

> If we think, after all, that the boat is a floating piece of space, a place without a place, that exists by itself, that is closed in on itself and at the same time is given over to the infinity of the sea and that, from port to port, from tack to tack, from brothel to brothel, it goes as far as the colonies in search of the most precious treasures they conceal in their gardens, you will understand why the boat...has been...the greatest reserve of the imagination. The ship is the heterotopia par excellence. In civilizations without boats, dreams dry up, espionage takes the place of adventure, and the police take the place of pirates.[111]

After the eviction of the Occupiers from Zuccotti Park, Occupy Wall Street may be seen as just this type of "place without a place": Occupy Wall Street is in Union Square, it is on the front steps of the CUNY Graduate Center, it is in Chicago, it is in over 80 countries on every continent, including Antarctica.[112] As with Foucault's heterotopic boat, Occupy Wall Street exists by itself, simultaneously enclosed within its chants of the "99%" and sparkle fingers, yet able to occur in an indefinite and infinite series of locations. In this way, Occupy Wall Street, and movements like it, are not only heterotopias that, through their consistent practice of liberty might themselves be able to liberate, but are also heterotopias *par excellence*. Occupy Wall Street is in ways a physical occupation, but, crucially, it is more an occupation of the political imagination: a small ship sailing somewhere across an infinite sea. Perhaps one day it will be widely acknowledged that without movements like Occupy Wall Street, dreams will dry up and espionage will continue to take the place of adventure. Or, perhaps, through these movements, we will be able to envision a world without police or pirates.

Notes

1. See http://whit537.org/2011/10/occupation.html.
2. See Anne Barnard, "Occupy Wall Street Meets Tahrir Square," *New York Times* (October 25, 2011). Indeed, after its occupation, Tahrir Square became known as "Liberation Square," much as Zuccotti Park became known as "Liberty Plaza Park." Indeed, "Liberty Plaza Park" had been the original name for Zuccotti Park, which was renamed in 2006 after John Zuccotti, the chairman of Brookfield Properties, the park's owner.
3. Just after midnight on November 15, nearly two months after the "occupation" began, the New York Police Department ordered everyone in Zuccotti Park to immediately leave, ostensibly for sanitary reasons so that the park might be cleaned. The police then forcibly removed anyone who remained, arresting over two hundred people, including several journalists. The police also removed, and in many cases damaged and destroyed, many of the items that had been in the park, from protestors' personal belongings to the books in the People's Library, the bicycle generators, and laptop computers that had been used by the organizers. See James Barron and Colin Moynihan, "City Reopens Park after Protesters Are Evicted," *NY Times* (November 15, 2011).
4. See for example, Lisa W. Foderaro, "Privately Owned Park, Open to the Public, May Make Its Own Rules," *New York Times* (Oct 13, 2011) (the article's first sentence begins "Zuccotti Park, the half-acre plaza in Lower Manhattan now synonymous with Occupy Wall Street"); Andrew Grossman, Alison Fox and Sean Gardiner, "Wall Street Protesters Evicted from Camp," *Wall Street Journal* (November 16, 2011) ("Occupy Wall Street activists grappled with the future of their movement after a police raid and a day of legal drama ended their hold on a Manhattan park that had become the symbolic center of the world-widep rotest").
5. For example, over 700 people were arrested on a march across the Brooklyn Bridge in early October of 2011, over 200 were arrested on the November 17 Day of Action, and over 100 were arrested on the 2011 New Year's Eve Occupy Action. Actions under the banner of "Occupy Wall Street" occurred throughout 2012 and 2013, including as part of May Day activities and celebrations of the one- and two-year anniversaries of the September 17 encampment. Occupy Wall Street also lead the relief efforts after Hurricane Sandy devastated parts of New York in 2012. See Alan Feuer, "Occupy Sandy: A Movement Moves to Relief," *NY Times* (November 12, 2012).

6. See Timothy Zick, "Space, Place and Speech: The Expressive Topography," 74 *George Washington Law Review* 439 (2006); Timothy Zick, "Speech and Spatial Tactics," 84 *Texas Law Review* 581 (2006); Thomas P. Crocker, "Displacing Dissent: The Role of 'Place' in First Amendment Jurisprudence," 75 *Fordham Law Review* 2587 (2007); Mark Kessler, "Free Speech Doctrine in American Political Culture: A Critical Legal Geography of Cultural Politics," 6 *Connecticut Public Interest Law Journal* 205 (2007).
7. See infra Section III.a.
8. See for example, Setha M. Low, and Denise Lawrence-Zuniga, "Locating Culture," in Setha M. Low and Denise Lawrence-Zuniga (eds.), *The Anthropology of Space and Place: Locating Culture* (Malden, MA : Blackwell Publication, 2003); Austin Sarat, Lawrence Douglas, and Martha Merrill Umphrey (eds.), *Law as Punishment / Law as Regulation* (Stanford, CA : Stanford Law Books, 2011); James Voorhies, "A Foucault Remix, or a Panoptic Gaze," in Voorhies, *Of Other Spaces* (n 1) (focusing on the disciplinary functions of space, even though this is an introduction to a series of works about Foucault's discussion of heterotopias and includes a reprint of *Of Other Spaces*, in which Foucault elaborates the concept of a "heterotopia"); but *cf* Stuart Elden, and Jeremy W. Crampton (eds.), *Space Knowledge and Power: Foucault and Geography* (Burlington, VT: Ashgate, 2007) (drawing largely upon lesser-known works of Foucault that stress the freedom made possible by space). It is interesting to note that scholars who utilize Foucault's arguments in the context of resistance within social movements tend not to discuss space. See for example, Bernard E. Harcourt, "Occupy Wall Street's 'Political Disobedience'," *New York Times*, October 13, 2011, http://opinionator.blogs.nytimes.com/2011/10/13/occupy-wall-streets-political-disobedience/; see also Bernard E. Harcourt, "Reflecting on the Subject: A Critique of the Social Influence Conception of Deterrence, the Broken Windows Theory, and Order-Maintenance Policing New York Style," 97 *Michigan Law Review* 291 (1998).
9. See infra Section III.a.
10. See infra notes 81–82.
11. Foucault, "Space, Knowledge and Power" (n 13), 252.
12. See for example, Foucault, "Of Other Spaces" (n 1); Michel Foucault, "Space, Knowledge and Power" (Christian Hubert, trans). In Paul Rabinow (ed.), *The Foucault Reader* (New York: Pantheon Books, 1984); Michel Foucault, *Discipline and Punish: The Birth of the Prison* (Alan Sheridan, trans.: New York: Random House, 1995), 195–210; Michel Foucault, "Questions on Geography," in Colin Gordon (ed. and trans.), *Power/Knowledge: Selected Interviews and Other Writings 1972–1977* (1980); Michel Foucault, "The Eye of Power," in Gordon, at 146 et seq; See also Elden and Crampton, *Space Knowledge and Power* (n 9), in which the authors translate into English for the first time several pieces of Foucault that deal explicitly with issues of space.
13. See for example, Michel Foucault, *Madness and Civilization: A History of Insanity in the Age of Reason*. (London: Routledge, 1967); Michel Foucault, *The Birth*

of the Clinic: An Archaeology of Medical Perception (London: Tavistock, 1976); Michel Foucault, "The Subject and Power." In Hubert Dreyfus and Paul Rabinow (eds.), *Michel Foucault: Beyond Structuralism and Hermeneutics* (Chicago: University of Chicago Press, 1982); Michel Foucault, *The Birth of Biopolitics— Lectures at the College de France, 1978–1979* (Basingstoke and New York: Palgrave Macmillan, 2008); Michel Foucault, *Society Must Be Defended: Lectures at the Collège De France, 1975–1976* (New York: Picador, 2003).

14. Foucault, "Eye of Power" (n 13), 149.
15. Foucault, *The Birth of the Clinic* (n 14), cited in Elden and Crampton, *Space Knowledge and Power* (n 9), 5.
16. See Michel Foucault, *Security, Territory, Population: Lectures at the College de France 1977–78* (Michel Senellart, ed., and Graham Burchell, trans, 2009), 16, in which Foucault defines "biopower" as "the set of mechanisms through which the basic biological features of the human species became the object of a political strategy, or a general strategy of power, or, in other words, how, starting from the eighteenth century, modern western societies took on board the fundamental biological fact that human beings are species."
17. See n 7 and accompanying text.
18. Zick, "Space, Place and Speech" (n 7) ("To the extent that 'place' enters constitutional discourse at all, it is as nothing more than a resource, a parcel of property, or an inert element of the expressive background." Ibid., 439); Crocker, "Displacing Dissent" (n 7) ("Under current First Amendment jurisprudence, public officials exercise increasingly effective means of displacing dissent through the regulation of place. By rendering dissent invisible, official control over the location of speech threatens a core, even romantic, value protected by the First Amendment." Ibid., 2587).
19. See Zick, "Space, Place and Speech" (n 7), 439.
20. See Zick, "Space, Place and Speech" (n 7), 439–440. For a discussion between public and non-public fora, see the United States Supreme Court case *Perry Educ. Ass'n v. Perry Local Educators' Ass'n*, 460 U.S. 37 (1983), 45–46.
21. Ibid., 440. Zick continues: "The analysis proceeds as if speech and spatiality are wholly separate and distinct elements; that places are mere abstract locations or properties to be categorized. So long as the government does not deny access to all places, the courts have reasoned, speakers have not been denied any constitutional right to communicate."
22. Ibid., (constitutional law on the First Amendment "does not adequately take into account that where a speaker or group of speakers is placed profoundly impacts expressive message, persuasive efficacy, participation, and symbolic meaning."); see also Zick, "Speech and Spatial Tactics" (n 7), 583 ("Purportedly neutral restrictions on place can and do cancel expressive and associative rights"); Crocker, "Displacing Dissent" (n 7) ("Architectural control of place, and the spaces that constitute a particular place, is an effective means of asserting state power without appearing to censor speech. Nonetheless, architecture is a means of control, as Michel Foucault famously argued." Ibid., 2600, n 97).

23. See Zick, "Speech and Spatial Tactics" (n 7), 581.
24. Ibid., 581–582; Crocker, "Displacing Dissent" (n 7), 2588–2589.
25. On the "freedom cage" in front of Federal Hall, see also John Del Signore, "Occupy Protesters Sue NYPD for Forcing Them into Free Speech Cages," *Gothamist* (April 30, 2012), *available at* http://gothamist.com/2012/04/30/second_ows_lawsuit.php. The articles regarding the use of nets, restrictions of sidewalks, metal barricades, and frozen zones are almost too numerous to list, and likely anyone living in New York City from September of 2011 to May of 2012 has seen more than one of these tactics being deployed. Anecdotally, I recently represented a medical worker accused of entering a "frozen zone" around a person being arrested, as he tried to reach the arrestee to determine if he needed medical attention. The reason that this case will likely be dismissed is that my client did not have notice of the frozen zone—not because we should not have "frozen zones" at political protests. In another incident, the NYPD actually locked Tompkins Square Park during its normal business hours in anticipation of an Occupy event there, locking around 50–100 people who were in the park at the time inside of it. See Allison Burtch, "NYPD Locks 100 People Inside Tompkins Square Park to Keep Occupy Out," *Animal New York* (May 23, 2012), *available at* http://www.animalnewyork.com/2012/nypd-locks-100-people-inside-tompkins-square-park-to-keep-occupy-out/.
26. Foucault, *Discipline and Punish* (n 13), 195–210.
27. Ibid., 199. For a further discussion of special branding and Foucault, see Voorhies, "A Foucault Remix, or a Panoptic Gaze" (n 9) ("To this point, [Foucault's major theoretical project] is about the function of human nature in our society and, in terms of space, the social relations generated by space as critical strands in vast, interconnected networks in which 'the subject is objectified by a process of division either within himself or from others.'" Ibid., 12, citing Michel Foucault, "The Subject and Power," in Dreyfus and Rabinow, *Michel Foucault: Beyond Structuralism and Hermeneutics* (n 14), 208.
28. Foucault, *Discipline and Punish* (n 13), 197.
29. Ibid., 219.
30. See Zick, "Speech and Spatial Tactics" (n 7), 629.
31. See Ibid., 628; see also Kessler, "Free Speech Doctrine in American Political Culture" (n 7), 210; Zick, "Space, Place and Speech: The Expressive Topography," (n 7), 475–479.
32. Voorhies, "A Foucault Remix, or a Panoptic Gaze" (n 9), 13. Voorhies continues: "In the end, space is about power; in the end, for Foucault, the social space of institutions is political." Ibid.
33. Documents released pursuant to Freedom of Information Act requests show that Occupy Wall Street was closely monitored by FBI Counterterrorism Agents, and the NYPD's counterterrorism squad was often seen at Zuccotti Park and other Occupy Wall Street events. See for example, Michael S. Schmidt and Colin Moynihan, "F.B.I. Counterterrorism Agents Monitored Occupy Movement, Records Show," *NY Times* (December 25, 2012).

34. Foucault, *Security, Territory, Population* (n 17), 26.
35. Ibid., 26.
36. See Lisa W. Foderaro, "Privately Owned Park, Open to the Public, May Make Its Own Rules," *NY Times* (October 13, 2011).
37. Foucault, *Security, Territory, Population* (n 17), 32.
38. Ibid.
39. Ibid., 33.
40. See for example, James Barron and Colin Moynihan, "City Reopens Park after Protesters Are Evicted," *NY Times* (November 15, 2011).
41. "Security will rely on a number of material givens." Foucault, *Security, Territory, Population* (n 17), 34.
42. "It is simply a matter of maximizing the positive elements, for which one provides the best possible circulation, and of minimizing what is risky and inconvenient, like theft and disease, *while knowing that they will never be completely suppressed*" (emphasis added) Foucault, *Security, Territory, Population* (n 17), 34.
43. "All these different functions of the town, some positive and others negative, will have to be built into the plan." Foucault, *Security, Territory, Population* (n 17), 34.
44. "The specific spaces of security refers then to a series of possible events; it refers to them temporal and the uncertain, which have to be inserted within a given space." Foucault, *Security, Territory, Population* (n 17), 35. The "milieu" is an important concept for Foucault in relation to security, especially the problems of "naturalness" that arises when considering the human species within an "artificial milieu." Ibid., 37. Foucault goes so far as to state the security is "a political technique that will be addressed to the milieu." Ibid., 38.
45. Quoted in James Barron, and Colin Moynihan, "City Reopens Park after Protesters Are Evicted," *NY Times* (November 15, 2011).
46. See n 32 and accompanying text.
47. It is an interesting question, and incredibly outside the scope of this paper, as to whether in these ways Occupy Wall Street may itself be considered a disciplinary society. Due to the mediating factors described below, in which Occupy Wall Street did not entirely separate, order, and brand itself, I believe that it should not be considered as such.
48. See for example, http://wearethe99percent.tumblr.com/.
49. See for example, Richard Kim, "We Are All Human Microphones Now," *The Nation* (October 3, 2011).
50. As but one small instance of the massive documentation of surveillance of Occupy Wall Street, see the leaked "Event Advisory Bulletin" of the New York Police Department, laying out in detail the plans for all Occupy-related May Day events, available at http://www.buzzfeed.com/rosiegray/new-york-police-department-anticipates-may-day-vio.
51. Indeed, this perhaps explains why the occupation remained for almost two months in Manhattan, and why arguments have been posed in the cases ensuing from the raid that, even after these months, neither Brookfield properties

nor the City of New York was able to deduce a legal reason for the eviction. See for example, New York Civil Liberties Union, "Memorandum of Law of *Amicus Curiae*," *People of the State of New York v. Ronnie Nunez, 2011* NY 0822981 (February 17, 2012), available at http://www.scribd.com/doc/81975361/NYCLU-Amicus-Brief.
52. Much of the media and scholarly attention to Occupy Wall Street has seen the mixture of messages as a negative aspect of the movement. See for example, Janean Chun, "Occupy Wall Street's Marketing Problem: Can Experts Help Solve an Identity Crisis?" *The Huffington Post* (November 17, 2011), available at http://www.huffingtonpost.com/2011/11/17/occupy-wall-streets-marketing-problem_n_1098422.html; but see Judith Butler, "So What Are the Demands? And Where Do They Go from Here?," 2 *tidal: Occupy Theory, Occupy Strategy* 8 (2012); Harcourt, "Occupy Wall Street's 'Political Disobedience'" (n 9) ("Occupy Wall Street, which identifies itself as a 'leaderless resistance movement with people of many…political persuasions,' is politically disobedient precisely in refusing to articulate policy demands or to embrace old ideologies.").
53. Zick, "Speech and Spatial Tactics" (n 7), 584.
54. Foucault, "Of Other Spaces" (n 1), 117. This text is based on a 1967 lecture by Foucault, and was first published in the French Journal *Architecture Mouvement Continuité* in 1984.
55. That Foucault relates space not only to power but also to resistance seems a natural result of his well-known position that "where there is power, there is resistance and yet this resistance is never in a position of exteriority in relation to power." Michel Foucault, *The History of Sexuality: An Introduction* (Robert Hurley, trans, London: Penguin, 1990), 95.
56. Foucault, "Of Other Spaces" (n 1), 117.
57. Ibid., 113.
58. Ibid., 113–114. Foucault notes that in Medieval times, this hierarchical arrangement of space both concerned the "real lives of men" and cosmological theory, in which the supercelestial was placed above the celestial, which was itself about the terrestrial. Ibid., 113.
59. Ibid., 114.
60. Ibid.
61. Ibid., 115.
62. Ibid.
63. Ibid. One example of a still-sanctified space is that of public bathrooms.
64. Ibid., 115–116. Foucault writes: "We do not live in a kind of void, inside of which we could place individuals and things."
65. Ibid., 116.
66. Ibid., 116–117 (emphasis added). Foucault also posits the existence of a third space, that of a "mirror" that negotiates between utopias and heterotopias, in that the mirror itself does exist (like a heterotopia), but what it reflects is unreal (like a utopia). Ibid., 117.
67. Foucault, "Of Other Spaces" (n 1), 118.

68. Ibid., 118. Foucault notes that some heterotopias are both of crisis and of deviation, such as nursing homes, where age is a crisis and leisure is a deviation in our society.
69. Ibid., 118–119.
70. Ibid., 120. Foucault here uses the precious example of the garden, which in its Oriental origins "had very deep and seemingly superimposed meanings." Foucault continues: "The traditional garden of the Persians was a sacred space that was supposed to bring together inside its rectangle four parts representing the four parts of the world...; and all the vegetation of the garden was supposed to come together in this space, in this sort of microcosm.... The garden has been a sort of happy, universalizing heterotopia since the beginning of antiquity." Ibid.
71. Foucault, "Of Other Spaces" (n 1), 121.
72. Ibid., 121. For example, the cemetery is a heterotopia in that it is only entered with the "heterochrony" of death, and "with this quasi-eternity in which her permanent lot is dissolution and disappearance."
73. Ibid.
74. Ibid. Foucault here also writes: "Such, for example, are the fairgrounds, these marvelous empty sites on the outskirts of cities that teem once or twice a year with stands, displays, heteroclite objects, wrestlers, snakewomen, fortune-tellers and so forth." Ibid 121–122.
75. Ibid., 122.
76. Ibid.
77. Ibid.
78. Ibid., 123.
79. Ibid., 124.
80. Ibid.
81. See Claire Tancons, "Occupy Wall Street: Carnival against Capital? Carnivalesque as Protest Sensibility," 30 *e-flux* (2011), available at http://www.e-flux.com/journal/occupy-wall-street-carnival-against-capital-carnivalesque-as-protest-sensibility/.
82. See Caroline Baillie, Jens Kabo, and John Reader (eds.), *Heterotopia: Alternative Pathways to Social Justice* (Hants, UK: Zero Books, 2013), 2.
83. For a further discussion of space and freedom in Foucault's work, see Elden and Crampton, *Space Knowledge and Power* (n 9), 9–10.
84. Foucault, "Space, Knowledge and Power" (n 13), 246.
85. Ibid., 245 (emphasis in original).
86. See Michel Foucault, "Force of Flight," cited in Elden and Crampton, *Space Knowledge and Power* (n 9), 7–8.
87. See for example, a study by Fordham University conducted from October 14 to October 18 of 2011, which found that, amidst many other points of diversity in viewpoints, "Ninety-seven percent of [the 300 Occupy Wall Street protesters] surveyed said they do not approve of how Congress is doing its job." Cited in

Melanie Eversley, "Occupy Wall Street Shows Diversity in Age, Politics," *USA Today* (November 1, 2011).
88. See n 66 and accompanying text.
89. See n 62 and accompanying text.
90. See for example, Jen Chung, "Protesters Having Sex, Losing Virginity at Occupy Wall Street," *Gothamist* (October 24, 2011), available at http://gothamist.com/2011/10/24/protesters_having_sex_losing_virgin.php.
91. See http://www.nycga.net/, the official website of the New York City General Assembly, which, as of March 23, 2012, lists 91 active working groups. See http://www.nycga.net/groups/.
92. See n 68 and accompanying text.
93. See n 88 and accompanying text.
94. See n 70 and accompanying text.
95. See the Declaration of the Occupation of New York City (adopted by the NYC General Assembly on Sep 29, 2011), available at http://www.nycga.net/resources/declaration/ (beginning: "As we gather together in solidarity to express a feeling of mass injustice, we must not lose sight of what brought us together. We write so that all people who feel wronged by the corporate forces of the world can know that we are your allies."); see also Harcourt, "Occupy Wall Street's 'Political Disobedience'" (n 9) ("Many of the voices at Occupy Wall Street accuse political ideology on both sides, on the side of free markets but also on the side of big government, for serving the few at the expense of the other 99 percent.").
96. See n 71 and accompanying text.
97. See n 88 and accompanying text.
98. Foucault, "Of Other Spaces" (n 1), 120; see also n 71 and accompanying text.
99. Ibid., 121; see also n 72 and accompanying text. Indeed, the origins of Occupy Wall Street are often attributed to the protesters having reached a "breaking point." See for example, Armin Rosen, "Occupy Wall Street Succeeds Where Bush-Era Peace Protests Failed," *The Atlantic* (October 20, 2011) (discussing an interview with an Occupier who "agrees that Occupy Wall Street reflects a burgeoning frustration with the social and economic status quo. 'It's good for people to realize they have limits,' she says. 'I'm relieved to see that people have a breaking point.'" Ibid.).
100. See http://peopleslibrary.wordpress.com/.
101. See www.occupywallstreet.org.
102. Readers may find images from the event in *tidal* magazine, which covered it in their December 2011 issue, http://tidalmag.org/pdf/tidal1_the-beginning-is-near.pdf.
103. See for example, Sarah Maslin Nir, "At Protest Central, Sleep, if You Can," *New York Times* (October 18, 2011) (visiting Zuccotti park as part of her "Nocturnalist" column, Nir refers to the encampment as a "hot new nightspot," reporting that "at 9:30 p.m., a couple was swing dancing to live accordion

accompaniment in the southeast corner. On the western end, three girls Hula-Hooped with illuminated hoops."). See also Tancons, "Occupy Wall Street" (n 82), arguing that "OWS might well be another Carnival Against Capital—a tactical re-territorialization of public space and political discourse, of social formation and cultural production, carried out as a concerted effort to regain democratic rights and liberties."). On a personal, anecdotal level, I recently represented a woman arrested for wearing rollerblades and butterfly wings at Zuccotti Park.

104. Foucault, "Of Other Spaces" (n 1), 122.
105. See n 78 and accompanying text. This criterion, itself, seems in conflict with the rest of Foucault's heterotopology, as it seems to preclude in some ways the positive effects that he sees heterotopias as able to have. See infra nn 82 and 83, and accompanying text.
106. See the *Hand Gestures Guide*, in the *General Assembly Guide*, available at http://www.nycga.net/resources/general-assembly-guide/.
107. See for example, Harcourt, "Occupy Wall Street's 'Political Disobedience'" (n 9) (arguing that "one way to understand the emerging disobedience is to see it as a refusal to engage these sorts of worn-out ideologies rooted in the Cold War ... [which] prop[ed] up the illusion of a free market and to legitimize the fantasy of less regulation.").
108. Note that this ideal is likely not an ordered, controlled one. See for example, Todd Gitlin, "Occupy Wall Street is Chaotic, Romantic and Utopian—and that's a Good Thing," *The New Republic* (October 18, 2011).
109. See n 85 and accompanying text.
110. On the continual and active nature of Occupy Wall Street, one need only consider its primary chant: "ALL DAY, ALL WEEK, OCCUPY WALL STREET."
111. Foucault, "Of Other Spaces" (n 1), 124.
112. See for example, Alan Taylor, "Occupy Wall Street Spreads Worldwide," *The Atlantic* (October 17, 2011), http://www.theatlantic.com/infocus/2011/10/occupy-wall-street-spreads-worldwide/100171/.

Bibliography

Butler, Judith. "So What Are the Demands? And Where Do they Go from Here?" *Tidal: Occupy Theory, Occupy Strategy* 2 (2012): 8–11.

Crocker, Thomas P. "Displacing Dissent: The Role of 'Place' in First Amendment Jurisprudence." *Fordham Law Review* 75 (2007): 2587–2639.

Elden, Stuart, and Jeremy W. Crampton (eds.). *Space, Knowledge and Power: Foucault and Geography*. Burlington, VT: Ashgate, 2007.

Foucault, Michel. *Madness and Civilization: A History of Insanity in the Age of Reason*. London: Routledge, 1967.

———. *The Birth of the Clinic: An Archaeology of Medical Perception*. London: Tavistock, 1976.

———. "The Subject and Power." In *Michel Foucault: Beyond Structuralism and Hermeneutics*, edited by Hubert Dreyfus and Paul Rabinow. Chicago: University of Chicago Press, 1982.

———. "Space, Knowledge and Power." In *The Foucault Reader*, edited by Paul Rabinow, translated by Christian Hubert. New York: Pantheon Books, 1984.

———. *The History of Sexuality: An Introduction*, translated by Robert Hurley. London: Penguin, 1990.

———. *Discipline and Punish: The Birth of the Prison*, translated by Alan Sheridan. New York: Random House, 1995.

———. *Society Must Be Defended: Lectures at the Collège De France, 1975–1976*. New York: Picador, 2003.

———. "Force of Flight." In *Space, Knowledge and Power: Foucault and Geography*, edited by Stuart Elden and Jeremy W. Crampton. Burlington, VT: Ashgate, 2007.

———. *The Birth of Biopolitics—Lectures at the College de France, 1978–1979*. Basingstoke and New York: Palgrave Macmillan, 2008.

———. "Of Other Spaces." In *Of Other Spaces*, edited by James Voorhies. Columbus, OH: Bureau for Open Culture, Columbus College of Art & Design, 2009a.

———. *Security, Territory, Population: Lectures at the College de France 1977–78*, edited by Michel Senellart, translated by Graham Burchell. New York: Picador, 2009b.

Harcourt, Bernard E. "Reflecting on the Subject: A Critique of the Social Influence Conception of Deterrence, the Broken Windows Theory, and order-Maintenance Policing New York Style." *Michigan Law Review* 97 (1998): 291–389.

———. "Occupy Wall Street's 'Political Disobedience'," *New York Times*, October 13, 2011. http://opinionator.blogs.nytimes.com/2011/10/13/occupy-wall-streets-political-disobedience/.

Kessler, Mark. "Free Speech Doctrine in American Political Culture: A Critical Legal Geography of Cultural Politics." *Connecticut Public Interest Law Journal* 6 (2007): 205–244.

Low, Setha M., and Denise Lawrence-Zuniga. "Locating Culture." In *The Anthropology of Space and Place: Locating Culture*, edited by Setha M. Low and Denise Lawrence-Zuniga. Malden, MA : Blackwell Publication, 2003.

New York Civil Liberties Union. "Memorandum of Law of *Amicus Curiae*." *People of the State of New York v. Ronnie Nunez*, 2011 NY 0822981, February 17, 2012. http://www.scribd.com/doc/81975361/NYCLU-Amicus-Brief.

Baillie, Caroline, Jens Kabo, and John Reader (eds.). *Heterotopia: Alternative Pathways to Social Justice*. Hants, UK: Zero Books, 2013.

Sarat, Austin, Lawrence Douglas, and Martha Merrill Umphrey (eds.). *Law as Punishment/Law as Regulation*. Stanford, CA: Stanford Law Books, 2011.

Tancons, Claire. "Occupy Wall Street: Carnival Against Capital? Carnivalesque as Protest Sensibility." *e-flux* 30 (2011). http://www.e-flux.com/journal/occupy-wall-street-carnival-against-capital-carnivalesque-as-protest-sensibility/.

Tidal: Occupy Theory, Occupy Strategy 1 (December 2011). http://tidalmag.org/pdf/tidal1_the-beginning-is-near.pdf.

Voorhies, James. "A Foucault Remix, or a Panoptic Gaze." In *Of Other Spaces*, edited by James Voorhies. Columbus, OH: Bureau for Open Culture, Columbus College of Art & Design, 2009.

Zick, Timothy. "Space, Place and Speech: The Expressive Topography." *George Washington Law Review* 74 (2006): 439–505.

Zick, Timothy. "Speech and Spatial Tactics." *Texas Law Review* 84 (2006): 581–651.

CHAPTER 7

Eating for the Future: Veganism and the Challenge of In Vitro Meat

Rasmus R. Simonsen

Introduction

Who hasn't fantasized about the end of the world? There is something about the destruction of all conventions and physical and ideological structures that fascinates the human imagination. David Agranoff's novel *The Vegan Revolution... With Zombies* sets out specifically to imagine what a world without the agricultural food industry would look like. In Agranoff's imagining, the vegan revolution is brought about by a regular zombie apocalypse. A new drug has been introduced that allows the meat industry to create what in the novel becomes known as Stress Free Food. Making animals immune to suffering, this drug has the unfortunate side effect of making any human who consumes Stress Free Meat into a zombie, craving a different meat altogether: human.[1] Presumably, the drug is administered not only to cows but also to all other animals in the agriculture industry, and, as such, vegans[2] are the only ones not affected by the zombie virus. As a consequence, vegans of the world are now left to fend off the hordes of rampaging and hungry undead that will not be satisfied by a Big Mac.

The novel is imbued with biopolitical references. At one point, the protagonist wonders why her boyfriend's particularly vile and obese neighbors were ever "allowed to breed."[3] Most of the novel details the beginning of the end and the battle to overcome the main obstacle standing in the way of the new meatless food utopia: zombies. In an obvious homage to George Romero's classic horror film, *Dawn of the*

Dead, the tagline of the book reads: "When there's no more meat in hell, the vegans will walk the earth..." After taking the reader through scenes of bloodshed and carnage, Agranoff's otherwise somewhat silly novel ends on a poignant note, as the author considers what the world would look like after the vegan revolution has run its course. At the end of the book, the aged protagonist, Dani—being the last person alive to have lived through what effectively became the dismantling of civilization as we know it—asks a group of school children if they "know the word vegan."[4] As they have never lived in a world of animal exploitation or factory farming, veganism is an obscure term to them. Agranoff writes: "[the children] looked at each other confused. Dani smiled. 'I suppose you wouldn't know that word, would you.' Dani closed her eyes and took in a deep breath. 'Good for you.'"[5]

Dani can die knowing that enduring change has been made—and all it took was a grand-scale zombie bloodbath that only the purest of vegans had a chance to live through. Dani's death, at the same time, also signals the end of veganism. When everybody is vegan, no one is. This leads us to ask the following questions: Can vegans be happy only after veganism as such has disappeared? What is the relationship between Agranoff's fictional Stress Free Meat and the dawn of in vitro meat, which is now in the beginning stages of production? What are the ethics involved, and how is the vegan consumer to respond to this new development—one that might make veganism irrelevant for the future of food? Subsequently, exploring the material, economic, and philosophical circumstances of in vitro meat, this chapter looks at how advances in food science affect the discussion of biopolitics and utopia. Throughout, references will be made to different contemporary thinkers and artists in order to understand how the tradition of utopian thinking influences ethical questions concerning the consumption of animals. In *Utopia in the Age of Globalization: Space, Representation, and the World-System*, Robert T. Tally defines utopia as "a fundamentally cartographic activity."[6] I will show how in vitro meat as a *potential* of cruelty-free consumption has been "mapped" onto vegan discourse in order to redefine the future of veganism. Finally, the chapter will return to the notion of future happiness as it relates to veganism and ethical consumerism.

Creating Artificial Meat

In 2008, People for the Ethical Treatment of Animals (PETA) offered a prize of $1 million to anyone able to make the first in vitro chicken meat and sell it to the public by June 30, 2012; they even provided their own recipe for fried "chicken," which contest participants were asked to follow. Since nobody submitted a piece of artificially grown chicken in time for the deadline, PETA decided to extend it to January 1, 2013, and then again to March 4, 2014. Despite the fact that to date no one has submitted a piece of in vitro chicken meat to PETA's contest, the organization's website still declares the contest "a smashing success!"[7] PETA does not claim this "success," but rather attributes it to scientists such as Dr. Mark Post who, ignoring PETA's taste for chicken, in 2013 produced the first in vitro hamburger. The burger was served up at an event in London and cost about \$325,000.[8] Although the production of in vitro meat products is clearly not economically viable at this point,[9] compared to contemporary standard meat production, "cultured meat involves approximately 7–45 percent lower energy use (only poultry has lower energy use), 78–96 percent lower GHG emissions, 99 percent lower land use, and 82–96 percent lower water use depending on the product compared."[10] In other words, the potential environmental benefits of replacing traditional meat with in vitro meat cannot be ignored.[11] But, how exactly is in vitro meat created?

In their rush to support this new advancement in food science, which they purport to be violence-free, PETA appears to have overlooked certain details of how in vitro meat is actually produced. The dream of "pure" meat being created *ex nihilo* in the lab is a fiction, since, as the creators of *The Tissue Culture and Art Project*, Oron Catts and Ionat Zurr explain in the essay "Towards a New Class of Being: The Extended Body," "important ingredients in (mainly) the nutrients provided to cells and tissues are derived from other living beings. One such ingredient is foetal [bovine] serum, which is used across the board to feed cultures of many cell types and origins."[12] Foetal bovine serum (FBS) is derived from fetuses that "are collected...from those animals deemed fit for human consumption," as William Siegel and Leland Foster point out, and as such FBS should be considered as "a by-product of the meat processing industry."[13] According to Humane Research Australia, "more than one million bovine fetuses [are harvested] annually."[14]

As a product, the strength of FBS is that, as Siegel and Leland note, "no other supplement has been found to provide the same degree and universality of cell growth stimulation. This cell growth stimulation comes from the abundance of blood-associated biochemicals responsible for the rapid cellular development inherent in fetal maturation."[15] FBS is habitually used in cell-based research, vaccine production, and now also in producing the first batches of in vitro meat. To produce FBS, blood is collected from bovine fetuses harvested from pregnant cows at the slaughterhouse. In order to ensure the quality and quantity of the product, the blood from the fetal heart must be collected while it is still beating, and the fetus expires as a result of exsanguination.[16] In other words, bioengineered meat is far from cruelty-free, as its production depends on the death of adult cows as well as their unborn offspring.

It is this matrix of life and death that generates the biopolitical impetus of in vitro meat. In *Animal Capital: Rendering Life in Biopolitical Times*, Nicole Shukin expands on Michel Foucault's by now well-known thesis on biopower in the first volume of *History of Sexuality*. According to Foucault, since the seventeenth century (or what he calls the "classical age"), the top-down hierarchy of sovereign power was gradually replaced by a "great bipolar technology—anatomic and biological, individualizing and specifying, directed toward the performances of the body, with attention to the processes of life—characteriz[ing] a power whose highest function was perhaps no longer to kill, but to invest life through and through."[17] Tracing the reevaluation and transformation of Foucault's original argument through the writings of Antonio Negri, Michael Hardt, and Giorgio Agamben, Shukin points to a blind spot in the biopolitical canon: in their rush to characterize the proliferation and technologies of invasive and subtle powers seeking to control entire human populations, these authors, Shukin argues, "are constrained by their reluctance to pursue power's effects beyond the production of human social and/or species life and into the zoopolitics of animal capital";[18] in response, Shukin's book crucially augments our conception of biopower to include "the ideological and affective functions of animal signs [and the] material institutions and technologies of speciesism."[19] Speciesism is a way of thinking that orders the world according to an artificial divide separating human from animal, and speciesism is therefore similar to racism and sexism. Deeming all animals inherently inferior to humans, speciesist thinking thus allows for the willful exploitation of animals for human gain, whether in terms of consumption, entertainment, or research.[20]

In a strange sense, laboratory-engineered meat seeks to erase the animal sign that nevertheless lies at the origin of its production. As Shukin has pointed out, biopolitics "operates through the power to hegemonize both the meaning and matter of life";[21] biopolitics, in other words, strives to create a system that values specific biological characteristics at the expense of other "inferior" elements related to sex, race, and species. In vitro science clearly shows this at work: the composite parts of in vitro meat combine, discursively, to *mean* free of cruelty. Even as the circumstances of producing FBS voids the cruelty-free designation, the discourse surrounding the production of in vitro meat gives shape to a hegemonic signification that ignores the actual material origin of in vitro meat. Invoking the second hegemonizing impetus of biopower, the *matter* of in vitro meat is hereby ordered according to a discourse of presence and absence. On the one hand, the end product is presented as a food object that, as if by magic, can boast all the same nutrients as regular meat, but retains none of the negative connotations that mar the latter—the animal origin, or sign, has been effaced. On the other hand, however, as a consequence of the discursive effacement of suffering, the hegemonizing signification of in vitro meat must include a forgetting of that small, yet vital, ingredient—FBS—that cannot be obtained in a way that would bypass the suffering or death of pregnant cows. As Thomas Lemke has pointed out, in the current era of bioengineering, life has largely been redefined as text: "The body is increasingly seen not as an organic substratum but as molecular software that can be read and rewritten."[22] Read textually, FBS can then be referred to as a dual-synecdoche, in the sense that it constitutes a (past) *part* of the dead fetus as well as a (future) *part* of in vitro meat. Conceptually, FBS straddles two temporalities (past/future) and two ontologies (life/death).

Contextualizing Shukin's theory of "rendering"—understood both in the sense of "economies of representation (the 'rendering' of an object on page, canvas, screen, etc.) and resource economies trafficking in animal remains (the business of recycling animal trimmings, bones, offal, and blood back into market metabolisms)"[23]—we see how in vitro meat has been rendered, in a *first* sense, as a phrase that designates the possibility of consuming meat with a clear conscience; whereas, in a *second* sense, the dead bovine fetus has been rendered into a liquid ingredient that guarantees the physical integrity of the cultured meat product. Extending Shukin's notion of rendering further, we should not forget the economic aspect involved in rendering fetal blood into FBS. Since FBS is a by-product of the beef industry, it appears logical that the pricing of FBS should follow the economic fluctuations of this industry. Depending on the supplier and the geographical factors involved, 500 ml of FBS can be had for about $200.[24] According to Siegel and Foster, the vagaries of the beef industry, while independent of the FBS market, nevertheless affect the pricing of FBS. The following events may influence the FBS market: (1) cattle sell-offs as a result of drought and harsh winters; (2) cattle retention due to ecological and economic changes in the agricultural industry; (3) dairy cow buy-outs to reduce milk production; and (4) shifts in milk and meat demand due to global weather or animal health conditions.[25] Furthermore, the geographical origin of the serum plays a big part in pricing. As some places in the world experience greater risk of viruses such as foot-and-mouth disease (FMD) and "mad cow disease" (BSE), the quality and pricing of FBS from these areas will be affected.

The bovine fetus, initially a wasteful remainder of the animal agriculture industry, is turned into a profitable serum that not so much supplements as completes the in vitro meat production process, thereby proving Shukin's point that the expansive "rendering industry scouts out an internal frontier ensuring capitalism will be able to continue its restless drive for economic expansion, training a new gaze inward on itself to cannibalize its own second nature."[26] If/when in vitro meat reaches an industrial scale of production, more and more fetuses will be needed in order to produce sufficient amounts of FBS. This kind of expansion, then, will either need to make available a segment of pregnant cows specifically for the harvesting of fetuses (cloning might also become a factor here), or an alternative will have to take the place of FBS. In time, FBS could require the establishment of a separate industry, which would mean that FBS could no longer be counted as simply a *by*-product of the beef industry.

Utopian Eating?

One basic tenet of utopia is freedom from scarcity; in this regard, the projected scope of in vitro meat can be considered utopian, as it aims at creating a food base that will no longer need to rely on the massive harvesting and processing of valuable resources such as water. An additional, if not primary, objective of in vitro meat is to remove pain and suffering from the equation by shifting the site of production from the slaughterhouse to the laboratory. So far, researchers have managed to move in the direction of completing each stated objective, without actually succeeding at either. This section deals primarily with the latter utopian aspect of in vitro meat: eliminating negative affect from the process and practice of meat consumption.

In order to address the ethical implications of the production and potential consumption of in vitro meat, in 2003 Oron Catts and Ionat Zurr constructed an art installation called *Disembodied Cuisine* in Nantes, France. As they describe it on their website, "the installation played on the notion of different cultural perceptions of what is edible and what is foul."[27] Two frog steaks were produced in vitro during the installation, but, as the artists are more than willing to point out, this "fake meat" was less than animal-friendly, as FBS was needed to give shape and texture to the frog meat. In general, Catts and Zurr are savvy to the inherent, yet productive, limitations of their art. In relation to their effort to produce "victimless leather" from cell materials, they note that, "[Victimless Leather] also presents an ambiguous and somewhat ironic take into the technological price our society will need to pay for achieving 'a victimless utopia'" ("Victimless Leather").[28] Catts and Zurr's work is important since, as Susan McHugh points out, it "give[s] the lie... to animal-friendly fantasies of real artificial meat, and, more importantly, [it] expose[s] the complex, multi-species agricultural and laboratory systems underpinning them."[29] In other words, Catts and Zurr's art aptly shows how in vitro meat, at its current stage, fails to meet the goal of utopian veganism as the complete absence of suffering.

It is namely the absence or negation of any "place" of suffering that is inherent to utopian veganism.[30] What we might also refer to as futural veganism attempts to modify or supplement the conventional understanding of utopia as a place exclusively concerned with "human needs and human flourishing."[31] Here we should note that, as Jean-Luc Nancy

Figure 7.1 "Disembodied Cuisine"— growing the steaks

Figure 7.2 "Disembodied Cuisine"

Figure 7.3 "Disembodied Cuisine"

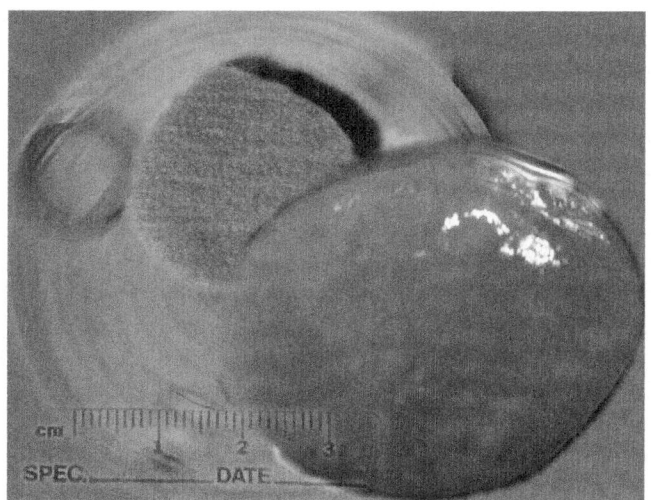

Figure 7.4 "Disembodied Cuisine"—dinner

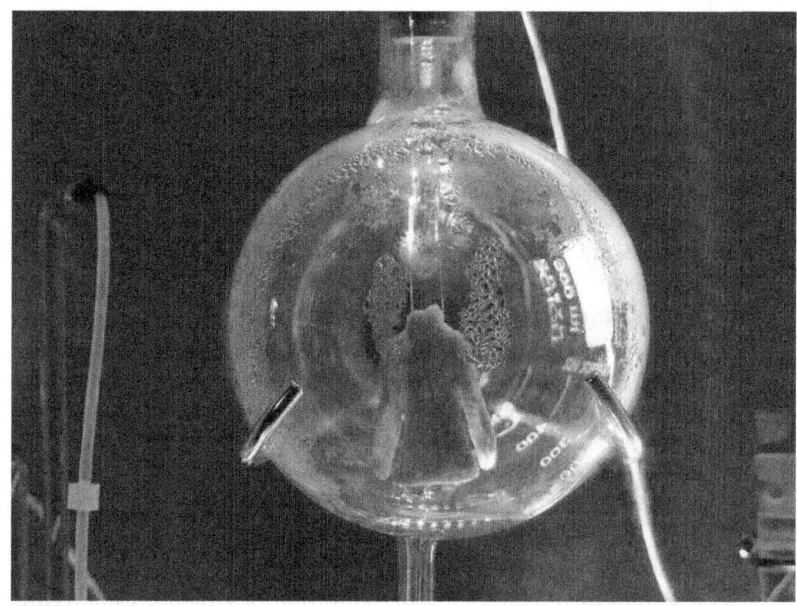

Figure 7.5 "Victimless Leather"

has pointed out, the word "utopia"—as conceived of by Thomas More in 1516, combining the Greek word τόπος ("place") and the negative prefix οὐ—"arose out of nothing in language as though by an act of creation *ex nihilo* [and is thus] given the task of designating a nothing-of-place, a non-place."[32] But how might one "conceive of a *topos* beyond all *topoi?*"[33] What we could call the meta-utopian art of Catts and Zurr taps into the "temptation"[34] central to the idea of utopia: the possibility of representing in language, or art, a point of completion where anticipation transforms into a concrete reality, unbounded by ideological or material strictures. However, the artists resist that temptation by pointing out the irony that at once undercuts and allows for an opening in their work to appear. *The Tissue Culture and Art Project* does not project a world without suffering; rather, by recognizing and discussing the presence of the negative elements that finally disrupt the dream of painless meat or victimless leather, Catts and Zurr construct a space that allows creator and spectator alike to both participate in and criticize what Gianni Vattimo has referred to as the fetishization of progress.[35] Setting the table for an ostensibly new relation to the consumption of animal products, Catts and Zurr with their installation have created a place masquerading as utopia; it is utopian in guise only, since, by relying on FBS to produce their meat, they secure an enduring link to the animal products industry.

Addressing the fluid boundaries of attraction and disgust (what is "edible" and what is "foul"), Catts and Zurr's installation is not based on the ethics of eating as such. It is therefore not surprising that McHugh should focus on Catts and Zurr's practice of "'scavenging leftovers' from research laboratories, as well as from food production, [in order to] create what they call 'semi-living sculptures.'"[36] McHugh is not immediately concerned with what their work might say about the ethics of eating, insofar as this touches on animal-free diets. In her article "Real Artificial: Tissue-cultured Meat, Genetically Modified Farm Animals, and Fictions," McHugh does not mention veganism, other than to say that "vegans should care more about the imperiled future of the world's pollinators than about what other people are eating."[37] To make up for McHugh's thin discussion of veganism, John Miller takes up the question of vegan identity politics in relation to the potential of "real fake meat" in his article, "In Vitro Meat: Power, Authenticity and Vegetarianism." He sees in vitro meat not as an alternative to current meat production, but rather as a "further symptom" of the "violent subjection of nonhuman animals."[38] As we have seen, the most obvious sign, or symptom, of violence underlying the production of in vitro meat concerns the harvesting of blood from bovine fetuses. According to Miller, since FBS is paramount to the engineering of artificial meat, it would be disingenuous to call it an "alternative" to regular meat.

Viewing in vitro meat as a symptom of already established processes of consumption and violence is important to consider further. We can explicate the symptomatic quality of in vitro meat by way of Slavoj Žižek's discussion of utopian socialism. In *The Sublime Object of Ideology*, Žižek discusses the utopianism of Marxian socialism, which he defines as the belief "that a society is possible in which the relations of exchange are universalized and production for the market predominates, but workers themselves none the less remain proprietors of their means of production and are therefore not exploited."[39] The utopian element here constitutes "the possibility of *a universality without its symptom*, without the point of exception functioning as its internal negation."[40] The "symptom" is the surplus value of labor, that which negates the "equitable exchange" between labor and capital. The fact that, in the modern system of wage labor, workers must "sell on the market their own labour instead of the products of their labour,"[41] thus becomes symptomatic of capitalism as such; simply put, it is the excess of this process that produces capital. When we transpose this insight onto the production of in vitro meat, from a vegan point of view, the addition of FBS constitutes the "internal negation" of the utopian potentiality of "painless" artificial meat; in other words, the simulacrum, or counterfeit meat product, fails exactly because it is not "fake" enough. FBS, in Žižekian parlance, is the symptom that disrupts the ethical motivation or outcome of in vitro meat.

Veganism itself inhabits an awkward position in relation to what we generally might call the "natural" order of things. Miller notes how the very artificiality of in vitro meat—which is no more unnatural than cheese or wine, we should add—is more disgusting to some than the violent reality of current practices in meat production; in this regard, he cites the "perceived opposition between authenticity and technicity" as the root cause of people's visceral reactions to in vitro meat.[42] Furthermore, he goes on to say that, "the popular cultural trope of a generalized science reversing the order of nature (as in the carnivalesque figuring of the ape as the master of man) has significant bearing on the initial reception of in vitro meat."[43] Commenting on George Orwell's novel *Coming Up for Air*, McHugh indeed suggests that fake meat is "nature's perversion in industrial society."[44] The following section will consider the binary of nature/artificiality and the extent to which

FBS can be said to inaugurate a post-animal basis for food production altogether.

Beyond Nature and Species

What is natural and what is not appears not only to be the question determining how one reacts to the prospect of eating in vitro meat, but, even more importantly, what it means to participate in the consumption of "real" meat. Carol Adams, in her influential work *The Sexual Politics of Meat*, deems the human consumption of meat unnatural, based on "the effects of high-fat diets on one's susceptibility to cancer and heart disease," and claims that, therefore, meat eating is "not consonant with the *human* body."[45] She then poses the following rhetorical question: "*If meat eating is natural, why do we not do it* naturally, *like the animals* [my emphasis],"[46] without using tools and heat to prepare the meat?[47] Of course, Adams's rhetorical question is disingenuous, insofar as were we to eat uncooked or raw meat, we would become susceptible to a number of adverse health effects. In Agranoff's novel, meat, especially bacon, is continually compared to the rear end of the animal from whence it came; thus, his preferred moniker for bacon is "ass shavings."[48] This crude allusion to the animal's behind—as an abject site of waste—in a most material way signals the unnaturalness of eating meat.

A return to "natural" consumption, then, appears to be the driving force behind Adams's and Agranoff's vision of veganism, whereas PETA is more than happy to jump on the in vitro wagon without properly ascertaining the actual "meaty" part of its production. Naturalness and artificiality emerge as the dividing terms of the debate. Looking beyond the materiality of in vitro meat, however, we should ask how the move toward artificially grown animal food products fits within a larger ideological or theoretical framework. Assuming that eventually researchers are able to make an animal-free growth serum for producing in vitro meat by the pound—made from either ground-up maitake mushrooms or blue-green algae, for instance[49]—when only the signifier remains of the "actual" animal we are consuming, what would this do to how we think about human and nonhuman relations?

Here, it will be helpful to consider what we might call the "post-animal" aspect of in vitro meat. As Nicola Jones points out in an article for *Nature*, establishing an enduring production of in vitro meat relies on creating an "immortal" stock of stem cells that would never have to be replenished; achieving this would mean that the species origin of the meat in question would no longer matter as such. Thinking beyond the limit of biological species is a stated aim of post-humanism as well. Very generally put, post-human theorists strive to abolish once and for all the binary thinking that is responsible for constructing and maintaining such divisions as human/animal, man/woman, hetero/homo, and so on. In *Journey to the End of the Species, 1: Guide to Singular Metamorphoses*, Thierry Bardini and Dominique Lestel specifically address the concept of the post-human in biological terms: "man [sic] today belongs to the species which abolishes species."[50] Positing a theory for the end of humanity—taking "the side of the machine against the animal"—Bardini and Lestel view the species as "a space of transition for the living being, who can thus live as several species throughout ... life";[51] "species," in other words, no longer will need to recognize death as a horizon of finitude. The ideological and ontological difference between humanity and animality would cease to matter at this point as well. Since we have reached a point where we are "rerout[ing] biological evolution to [our] advantage," the future maxim of post-humanity should read: *"I can be what I am prepared to accept."*[52] Immortality, in this instance, would be the final point of acceptance, that which would finally erode one of the chief concerns of the human imagination: the notion of an ending. Would in vitro meat not be a fitting component in an immortal post-human diet? After all, the post-human project shares with in vitro meat the same goal: achieving immortality on the cellular level. *I can* [eat] *what I am prepared to accept.*

Ironically, for Lestel, if not Bardini, species-less or post-animal food is not something he is prepared to accept, judging from his 2011 book *Apologie du carnivore* [In Defense of Carnivores]. In this book, Lestel chastises vegetarians for neglecting or disavowing their own animal nature, whereas he praises carnivores for insisting on ingesting other animals in order to maintain a level of intimacy with their own animal selves. Animality in this optic is equal to carnivorism, which quite obviously disregards the fact that the most commonly consumed animals are herbivores or omnivores. Lestel's "apology" of meat eaters is strangely romantic, as he claims that, somehow, by ingesting animals, one is ingesting "wildness." Carnivores, "instead of being limited to their ambition to want to live in peace with [animals], ... want to blend [*melanger*] with them" by eating them.[53] He goes so far as to say that eating meat is an ethical duty: "I call this the *carnivorous imperative.*"[54] Thus, Lestel's logic goes, consuming animals is akin to consuming the distance between the categories of "human" and "animal." By quite literally effacing animals (eating them), Lestel imagines that we can efface the ontological priority of the human. Clearly, Lestel's carnivores would lament the replacing of real meat with its in vitro counterpart, as the distance between the consumed animal and the consuming human appears complete at this stage; but not quite, since, as we have seen, for in vitro meat to take shape, animal ingredients must still be added to the process. Since FBS connects in vitro meat to the visceral element of the animal agricultural industry, we can only consider in vitro meat post-animal on the discursive level.

However, despite Lestel's argument to the contrary, we should note that the logic behind in vitro meat is not fundamentally different from that of carnivorism: the consumption of meat relies on, if not a forgetting, then an onto-ethical distancing by which the origin of meat, the animal, disappears or is obscured. What is left on the plate is neither wild nor brimming with the potential of becoming-animal (unless Lestel wants to make an argument for becoming-corpse).

Vegan or meat eater, there seems to be a race to judge the other side as being either unnatural or dogmatic. However, Lestel's carnivorism can agree with veganism on one point: industrial farming is fundamentally unsound. The reality of the industrial meat production complex of today can seem reminiscent of dystopian fantasy, and it is perhaps for this reason that much vegan discourse centers on utopian imagery as a counterpoint to the current system of production. This could also be the reason why organizations like PETA have been so quick to embrace in vitro meat as the answer to animal suffering and to the negative feelings associated with veganism. However, as Miller points out, in vitro meat would most likely not replace regular meat, but rather it would seek out its own niche in the market place. Accordingly, it does not seem likely that we will find ourselves in a situation like what Agranoff describes in the final pages of his novel, where the question of veganism has become moot. We thus find ourselves revisiting the question of veganism's affective dimension, and how this relates to its ethical component.

Happy Meals?

In order to approach an alternative to the fantasy of victim-less meat, and the idea that eating meat itself can somehow be disassociated from any notion of violence, this chapter ends by considering the concept of happiness that was alluded to at the beginning, since happiness (or the absence thereof), as it were, appears to be inseparable from the discussion of veganism and the future of food. If, according to Sara Ahmed, "happiness is often described as 'what' we aim for, as an end point,"[55] when there is no end in sight is it then possible to be happy? As it seems impossible at present to glimpse an end to animal suffering, can there be such a thing as happy veganism—without resorting to naïve utopianism, that is?

Perhaps the goal of veganism should not be to rid the world of negative affect. As Ahmed claims, "unhappiness [can be] a form of political action: the act of saying no or of pointing out injuries as an ongoing present affirms something, right from the beginning."[56] Specifically building on this, the vegan perspective is apt to make us look at how eating meat affects the place we occupy in the world. In this way, unhappiness becomes a form of resistance, understood in Foucauldian terms. Generally speaking, beginning with *The History of Sexuality*, Foucault sees power relations as mobile and changeable; resistance is thus part and parcel of power. In this regard, Foucault emphasizes the creative element of power. In our current context, this means that unhappiness is not simply the negative obverse of happiness; rather, unhappiness engenders a range of possibilities and points of exchange that challenge the basis of what it means to be happy in the first place. In this final section, then, I will suggest that we view veganism as fundamentally unhappy, in the sense that negative affect becomes a means of resistance.

"Happy" food choices are linked to a history and teleology of consumption. The tradition of what and how we eat immediately connects us to images of the past and fantasies of the future. It is interesting to note here that in the history of the Western family nexus, meat was typically considered the husband's food, while children and wives up until recently had to content themselves with mostly vegetables at the dinner table. In his 1979 course on biopolitics at the Collège de France, Foucault explains how the market economy, the rule of supply and demand, has been exported to social spheres outside the economic. Advancing "a sort of economic analysis of the non-economic,"[57] he suggests that the mother-child relationship, for example, can be measured in time as the production of human capital:

> concretely characterized by the time spent by the mother with the child, the quality of the care she gives, the affection she shows, the vigilance with which she follows its development, its education, and not only its scholastic but also its physical progress, the way in which she not only

gives it food but also imparts a particular style to eating patterns, and the relationship she has with its eating.[58]

Consequently, what Foucault calls human capital in this regard—as the potential of each human being to fill out a future place in the modern labor matrix, to "produce an income"[59]—is not only tied up with the child's intellectual development, but also its physical and social memorization of distinct "eating patterns." For the mother, the bottom line is measured in "psychical income": "She will have the satisfaction a mother gets from giving the child care and attention in seeing that she has in fact been successful."[60] Eating patterns, therefore, to a large extent, follow gendered lines.

Part of the maturation process, especially for males, is directed toward the future promise of meat consumption and what it leads to: a place at the end of the proverbial table. Meat, in this instance, becomes "a means to happiness," since, as Ahmed points out, "things become good, or acquire their value as goods, insofar as they point toward happiness."[61] It is clear from this that happiness is a horizon that we become directed toward through our interaction with certain objects. Here is Ahmed again: "For a life to count as a good life, it must take on the direction promised as a social good, which means imagining one's futurity in terms of reaching certain points along a life course."[62] But what happens when the direction is altered, when the path toward a proscribed form of happiness encounters an obstacle? Due to the vast changes that farming and animal rearing have undergone in the last century or so, the kind of meat consumed today is not the same as that of yesteryear; nor have the animals been treated in a similar manner. Be that as it may, contemporary meat-eating culture still takes for granted, Adams argues, "the normativeness and centrality of their activity."[63] The advent of in vitro meat is not likely to change this, as new technologies are simply aimed at ensuring the prominence of meat as we move into the future.

Nonetheless, in recent years, we have witnessed a push toward so-called ethical consumerism. At the forefront of ethical consumerism, we find the American grocery chain, Whole Foods Market. Since 1980, as Michael Serazio points out, Whole Foods "has grown to become the world's leading retailer of natural and organic foods,"[64] with close to 300 stores in North America and Britain. Not exclusively vegan, Whole Foods nevertheless presents the modern vegan with an array of processed soy products and other specialty items, which had hitherto been relegated to a niche market. But the sheer abundance of product (not to say, lifestyle) choices found at Whole Foods (their website describes shopping at Whole Foods as a "transcendent experience") has infused ethical consumption with bourgeois spectacle. As Serazio puts it,

> the Whole Foods experience—titillating every sense at bourgeois prices—... may presage the luxurious future of the counterculture imagination. It is the grocery store functioning as a kind of adult playland—a posh, almost lurid indulgence of food flavor, texture, and scent.... It tries to appear as the utopia of ethos groceries, the temple of food on one's path to consumption nirvana.[65]

The example of their "US$15 million, 80,000 square foot Austin flagship," completed in 2007,[66] perhaps most clearly emphasizes the fact that the Whole Foods brand of utopianism is located in a distinctive space-time, cut off from any motivation or action other than the impulse to shop; but shopping itself at Whole Foods is almost secondary to what Betsy Spethmann has termed the "linger longer" effect.[67] Consumer time is slow time. Subsequently, the kind of veganism that Whole Foods stocks in their stores does not seem to differ radically from conventional consumerism.

When ethical consumption becomes too comfortable, bordering on the mundane, what is left of the original impetus to make a change in the world? For veganism to remain effective, it must continue to insist on disrupting or deviating from the central social directionality of consuming animal products that Whole Foods and the new herbivores with Michael Pollan at the helm have done nothing to radically challenge. Veganism allows us to confront, and even cultivate, a certain kind of productive discomfort. And in this regard, it is more helpful to think about a kind of utopian veganism that stresses "process" (movement) over "product" (destination).[68]

Discomfort and deviation, as their prefixes indicate, are negative processes. Discomfiture belongs to the group of affects that is typically viewed as what Sianne Ngai calls "syntactically" negative, which points to affective states that are "organized by trajectories of repulsion rather than attraction."[69] Deviation, while not itself an affect, seeks out what is rendered repulsive or at odds with the norm. Veganism, as an attitude, is directed away from human carnivorism, and, as such, it expresses a discomfort with meat and other animal products. This discomfort does not correspond to what Lestel calls vegetarians' inherent anti-animalism; on the contrary, veganism identifies with, or attempts to project, the suffering that animals experience in the global agricultural industry.

But deviation as a turn away from meat is not necessarily a negative process; after all, turning away from one thing or path can propel us toward a different outcome by different means.

A century ago, meat was considerably more expensive and less frequently part of the dinner meal; cutting out meat and other animal products, in this way, might for older generations be considered a step backward, part of a certain social and economic regression. Here we should remember that the meat with every meal tradition is class-based and part of a specifically Western bias. Even today, many socioeconomic groups around the world only rarely partake of meat. In Western societies, not least of all the US, different groups consume different kinds of meat. The poorest members of society tend to eat more processed meats, and the hierarchy of society, to an extent, is reflected in the species, preparation, and cut of meat that a person consumes.

However, regardless of class, in Western, meat-dominated diets, the happiness space of the family is in danger of being contaminated by negative affect when veganism is introduced; in a sense, we can say that the space of the dinner table has been compromised by the addition of a "foreign" element. The social psychologist, Kurt Lewin, has written that the food habits of a given group "are the result of a multitude of forces. Some forces support each other, some oppose each other. Some are driving forces, others restraining forces. Like the velocity of a river, the actual conduct of a group depends upon the level...at which these conflicting forces reach a state of equilibrium."[70] What is allowed to pass, what is considered a "good" food object, depends on the ideology of the group, in other words. The threshold to the dinner table is guarded by varying beliefs and values,[71] and as the equilibrium of the family dynamic is upset, the temporal horizon of happiness as it pertains to the *telos* of the family can come to seem uncertain in turn. The genealogical affinity and continuity between past, present, and future no longer appear self-evident. And to refuse the preordained direction that the family as *telos* represents is at the same time "to cause unhappiness," as Ahmed intones, "whether or not you feel the unhappiness you are assumed to cause."[72]

Feelings of discontent can fuel the engine of critical inquiry. As this chapter has shown, allowing in vitro meat to pass the threshold of human consumption reveals the industrial and ideological processes involved in rendering foetal bovine blood into FBS. The utopianism of bioengineered meat clashes with its own material actuality. Moreover, as a cultural, scientific, and capitalistic "force," in Lewinian terms, this new constellation of meat product certainly clashes with veganism as well. However eager we might be to inhabit a future free of suffering and unnecessary death, we cannot bypass the facts and circumstances of our present scientific capabilities of achieving this. Cruelty-free meat may simply be another element of the fantasy that humanity will ever be able to dwell with and among other species equitably.

Notes

1. Similarly, Chris Cooney's cookbook, *The Vegan Zombie: Cook and Survive* (based on the eponymous YouTube show), imagines a scenario in which consuming animal products becomes the leading cause of zombification; in turn, this scenario (included in the book as a comic book narrative) comes to form the basis for developing "zombie-free dishes" (9), which are simply an array of vegan recipes with names such as "The Vegan Zomblette" (29). Generally speaking, the recipes do not seem suitable for the kind of migratory lifestyle that survivors take up in most zombie dramas (cf. *The Walking Dead*, for example).
2. Vegans abstain from consuming *all* animal products, including any type of meat, fish, dairy, eggs, and often honey. Additionally, most vegans will not wear leather or wool; some also object to wearing silk.
3. David Agranoff, *The Vegan Revolution... With Zombies* (Portland, OR: Deadite Press, 2010), 23.
4. Ibid., 153.
5. Ibid., 154.
6. Robert T. Tally, *Utopia in the Age of Globalization: Space, Representation, and the World System* (Houndmills, Basingstoke, and Hampshire, UK: Palgrave Macmillan, 2013), 5.
7. See http://www.peta.org/features/vitro-meat-contest/#ixzz33ydu4G4v.
8. Henry Fountain, "Building a $325,000 Burger," *The New York Times*, May 12, 2013, http://www.nytimes.com/2013/05/14/science/engineering-the-325000-in-vitro-burger.html?pagewanted=all&_r=0.
9. Hanna Tuomisto and M. Joost Teixeira de Mattos note that "about $160 million investments in research are needed for commercializing the production" (Tuomisto and Teixeira de Mattos, "Environmental Impacts of Cultured Meat Production," *Environmental Science & Technology* 45 [2011]: 6117).
10. Ibid., 6117.
11. As it stands, however, since waiting does nothing, the best way to influence the environment positively through consumption is by going vegan. As Vasile Stanescu has pointed out, "shifting from beef to vegetables for even a single day a week" would significantly impact one's carbon footprint in a positive way ("'Green' Eggs and Ham?': The Myth of Sustainable Meat and the Danger of the Local," in *Critical Theory and Animal Liberation*, edited by John Sanbonmatsu [Plymouth, UK: Rowman and Littlefield, 2011], 242).

12. Oron Catts, and Ionat Zurr, "Towards a New Class of Being: The Extended body." *Intelligent Agent* 6.2 (2006): 2, November 26, 2014, http://www.intelligentagent.com/archive/Vol6_No2_transvergence_cattszurr.htm.
13. William Siegel, and Leland Foster, "Fetal Bovine Serum: The Impact of Geography," *BioProcessing Journal* 12.3 (Fall, 2013): 28. November 26, 2014, http://dx.doi.org/10.12665/J123.Siegel.
14. Shatha Hamade, "Use of Fetal Calf Serum," *Humane Research Australia*, November 26, 2014, http://www.humaneresearch.org.au/campaigns/fetal_calf_serum.
15. Siegel and Foster, "Fetal Bovine Serum," 28.
16. D. J. Mellor, and N. G. Gregory, "Responsiveness, Behavioural Arousal and Awareness in Fetal and Newborn Lambs: Experimental, Practical and Therapeutic Implications," *New Zealand Veterinary Journal* 51.1 (2003): 3, http://www.sciquest.org.nz/node/36323.
17. Nicole Shukin, *Animal Capital: Rendering Life in Biopolitical Times* (Minneapolis:University of Minnesota Press, 2009), 139.
18. Ibid., 11.
19. Ibid.
20. For a more detailed examination of speciesism, see Peter Singer's seminal work *Animal Liberation: The Definite Classic of the Animal Movement* (New York: Harper Perennial, 2009).
21. Shukin, *Animal Capital*, 20.
22. Thomas Lemke, *Biopolitics: An Advanced Introduction*, translated by Eric Frederick Trump (New York: New York University Press, 2011), 93.
23. Shukin, *Animal Capital*, 21–22.
24. For example, 500 ml of Fetal Bovine Serum is priced at $215.00 on RMBIO's website (http://www.rmbio.com/fetal-bovine-serum). RMBIO is a premier producer of "cell culture supplements."
25. Siegel and Foster, "Fetal Bovine Serum," 29.
26. Shukin, *Animal Capital*, 68.
27. Catts and Zurr, "Disembodied Cuisine," *The Tissue Culture and Art Project*, November 26, 2014, http://tcaproject.org/disembodied-cuisine.
28. Catts and Zurr, "Victimless Leather: A Prototype of Stitch-less Jacket Grown in a Technoscientific 'Body,'" *The Tissue Culture and Art Project*, November 26, 2014, http://tcaproject.org/vl/.
29. Susan McHugh, "Real Artificial: Tissue-cultured Meat, Genetically Modified Farm Animals and Fictions," *Configurations* 18.1–2 (Winter, 2010): 189–190.
30. Patricia Vieira and Michael Marder point out that the thinking of utopia "from the outset announces a certain negation of place, *topos*" (*Existential Utopia: New Perspectives on Utopian Thought*, edited by Patricia Vieira and Michael Marder [New York: Continuum, 2012], ix.) The *topoi* that veganism hopes to negate would include slaughterhouses, dairy farms, egg hatcheries, fur farms, and animal-testing labs.
31. Ruth Levitas, *Utopia as Method: The Imaginary Reconstruction of Society* (Houndmills, Basingstoke, and Hampshire, UK: Palgrave Macmillan, 2013), xi.

32. Jean-Luc Nancy, "In Place of Utopia," in *Existential Utopia: New Perspectives on Utopian Thought*, edited by Patricia Vieira and Michael Marder (New York: Continuum, 2012), 3.
33. Ibid., 11.
34. Ibid., 10.
35. Gianni Vattimo, "Utopia, Counter-Utopia, Irony," in *Existential Utopia: New Perspectives on Utopian Thought*, edited by Patricia Vieira and Michael Marder (New York: Continuum, 2012), 20.
36. McHugh, "Real Artificial," 188.
37. Ibid., 196.
38. John Miller, "*In vitro* Meat: Power, Authenticity and Vegetarianism," *Journal for Critical Animal Studies* 10.4 (2012): 45, November 26, 2014, http://www.criticalanimalstudies.org/volume-10-issue-4-2012/.
39. Slavoj Žižek, *The Sublime Object of Ideology* (London: Verso, 1989), 18.
40. Ibid., 18.
41. Ibid., 17.
42. Miller, "*In vitro* Meat," 46.
43. Ibid., 45.
44. McHugh, "Real Artificial," 181.
45. Carol Adams, *The Sexual Politics of Meat: A Feminist-Vegetarian Critical Theory* (New York: Continuum, 2010), 196, 204.
46. Ibid., 198.
47. For a sustained discussion of Adams's position on the "unnaturalness" of eating meat, see Rasmus Simonsen, "A Queer Vegan Manifesto," *Journal for Critical Animal Studies* 10.3 (2012): 51–80, November 26, 2014, http://www.critical-animalstudies.org/volume-10-issue-3-2012/.
48. Agranoff, *The Vegan Revolution... With Zombies*, 18.
49. See Nicola Jones, "Food: A Taste of Things to Come?," *Nature* 468 (December 8, 2010): 752–753, November 26, 2014, http://www.nature.com/news/2010/101208/full/468752a.html.
50. Thierry Bardini, and Dominique Lestel, *Journey to the End of the Species, 1: Guide to Singular Metamorphoses* (Paris: Dis Voir, 2010), 9.
51. Ibid., 6, 8.
52. Ibid., 10.
53. Dominique Lestel, *Apologie du carnivore* (Paris: Fayard, 2011), 16. I am indebted to Andrew Weiss for providing this translation.
54. Ibid., 15.
55. Sara Ahmed, "Happy Futures, Perhaps," *Queer Times, Queer Becomings*, edited by E. L. McCallum and Mikko Tuhkanen (Albany: State University of New York Press, 2011), 163.
56. Sara Ahmed, *The Promise of Happiness* (Durham: Duke University Press, 2010), 207.
57. Michel Foucault, *The Birth of Biopolitics: Lectures at the Collège de France*, edited by Michel Senellart, translated by Graham Burchell (Houndmills, Basingstoke, and Hampshire, UK: Palgrave Macmillan, 2008), 243.

58. Ibid., 243–244.
59. Ibid., 244.
60. Ibid.
61. Ahmed, "Happy Futures, Perhaps," 163.
62. Ibid., 164.
63. Adams, *The Sexual Politics of Meat*, 201.
64. Michael Serazio, "Ethos Groceries and Countercultural Appetites: Consuming Memory in Whole Foods' Brand Utopia," *The Journal of Popular Culture* 44.1 (2011): 159, November 26, 2014, http://digitalcommons.fairfield.edu/communications-facultypubs/11.
65. Ibid., 174.
66. Ibid., 173.
67. Ibid.
68. Ronald Bogue, "Deleuze and Guattari and the Future of Politics: Science Fiction, Protocols and the People to Come," *Delezue Studies* 5 (2011): 87, November 26, 2014, http://www.euppublishing.com/doi/full/10.3366/dls.2011.0038.
69. Sianne Ngai, *Ugly Feelings* (Cambridge, MA: Harvard University Press, 2005), 11.
70. Kurt Lewin, *Field Theory in Social Science*, edited by Dorwin Cartwright (Westport, CT: Greenwood Press, 1975), 173.
71. Ibid., 186.
72. Ahmed, "Happy Futures, Perhaps," 165.

Bibliography

Adams, Carol. *The Sexual Politics of Meat: A Feminist-Vegetarian Critical Theory*. New York: Continuum, 2010.

Agranoff, David. *The Vegan Revolution... With Zombies*. Portland, OR: Deadite Press, 2010.

Ahmed, Sara. "Happy Futures, Perhaps." In *Queer Times, Queer Becomings*, edited by E. L. McCallum and Mikko Tuhkanen. Albany, NY: State University of New York Press, 2011.

———. *The Promise of Happiness*. Durham: Duke University Press, 2010.

Bardini, Thierry, and Dominique Lestel. *Journey to the End of the Species, 1: Guide to Singular Metamorphoses*. Paris: Dis Voir, 2010.

Bogue, Ronald. "Deleuze and Guattari and the Future of Politics: Science Fiction, Protocols and the People to Come." *Delezue Studies* 5 (2011): 77–97. http://www.euppublishing.com/doi/full/10.3366/dls.2011.0038.

Catts, Oron, and Ionat Zurr. "Towards a New Class of Being: The Extended Body." *Intelligent Agent* 6.2 (2006), November 26, 2014. http://www.intelligentagent.com/archive/Vol6_No2_transvergence_cattszurr.htm.

———. "Victimless Leather: A Prototype of Stitch-less Jacket Grown in a Technoscientific 'Body'." *The Tissue Culture and Art Project*, November 26, 2014. http://tcaproject.org/vl/.

———. "Disembodied Cuisine." *The Tissue Culture and Art Project*, November 26, 2014. http://tcaproject.org/disembodied-cuisine/.

Cooney, Chris, and Jon Tedd. *The Vegan Zombie: Cook and Survive!* Illustrated by Rob Kramer. Self-published.

Foucault, Michel. *The Birth of Biopolitics: Lectures at the Collège de France*, edited by Michel Senellart, translated by Graham Burchell. Houndmills, Basingstoke, and Hampshire, UK: Palgrave Macmillan, 2008.

———. *The History of Sexuality: Volume 1*, translated by Robert Hurley. New York: Pantheon Books, 1978.

Fountain, Henry. "Building a $325,000 Burger." *The New York Times*, May 12, 2013. http://www.nytimes.com/2013/05/14/science/engineering-the-325000-in-vitro-burger.html.

Hamade, Shatha. "Use of Fetal Calf Serum." *Humane Research Australia*, November 26, 2014. http://www.humaneresearch.org.au/campaigns/fetal_calf_serum.

Jones, Nicola. "Food: A Taste of Things to Come?" *Nature* 468 (December 8, 2010): 752–753, November 26, 2014. http://www.nature.com/news/2010/101208/full/468752a.html.

Lemke, Thomas. *Biopolitics: An Advanced Introduction*, translated by Eric Frederick Trump. New York: New York University Press, 2011.

Lestel, Dominique. *Apologie du Carnivore*. Paris: Fayard, 2011.

Levitas, Ruth. *Utopia as Method: The Imaginary Reconstruction of Society*. Houndmills, Basingstoke, and Hampshire, UK: Palgrave Macmillan, 2013.

Lewin, Kurt. *Field Theory in Social Science*, edited by Dorwin Cartwright. Westport, CT: Greenwood Press, 1975.

Marder, Michael, and Patricia Vieira. "Introduction." In *Existential Utopia: New Perspectives on Utopian Thought*, edited by Patricia Vieira and Michael Marder. New York: Continuum, 2012.

McHugh, Susan. "Real Artificial: Tissue-cultured Meat, Genetically Modified Farm Animals and Fictions." *Configurations* 18.1–2 (Winter, 2010): 181–197.

Mellor, D. J., and N. G. Gregory. "Responsiveness, Behavioural Arousal and Awareness in Fetal and Newborn Lambs: Experimental, Practical and Therapeutic Implications." *New Zealand Veterinary Journal* 51.1 (2003): 2–13. http://www.sciquest.org.nz/node/36323.

Miller, John. "*In Vitro* Meat: Power, Authenticity and Vegetarianism." *Journal for Critical Animal Studies* 10. 4 (2012): 41–63, November 26, 2014. http://www.criticalanimalstudies.org/volume-10-issue-4-2012/.

Nancy, Jean-Luc. "In Place of Utopia." In *Existential Utopia: New Perspectives on Utopian Thought*, Edited by Patricia Vieira and Michael Marder. New York: Continuum, 2012.

Newkirk, Ingrid E. "*In Vitro* Meat Prize Deadline Extended." 25 June 2012. *People for the Ethical Treatment of Animals*. 2014, November 26, 2014. http://www.peta.org/blog/vitro-meat-prize-deadline-extended/.

Ngai, Sianne. *Ugly Feelings*. Cambridge, MA: Harvard University Press, 2005.

"PETA's 'In Vitro' Chicken Contest." *People for the Ethical Treatment of Animals*, n.d. November 26, 2014. http://www.peta.org/features/vitro-meat-contest/.

Serazio, Michael. "Ethos Groceries and Countercultural Appetites: Consuming Memory in Whole Foods' Brand Utopia." *The Journal of Popular Culture* 44.1 (2011): 158–177, November 26, 2014. http://digitalcommons.fairfield.edu/communications-facultypubs/11.

Shukin, Nicole. *Animal Capital: Rendering Life in Biopolitical Times.* Minneapolis: University of Minnesota Press, 2009.

Siegel, William and Leland Foster. "Fetal Bovine Serum: The Impact of Geography." *BioProcessing Journal* 12.3 (Fall 2013): 28–30, November 26, 2014. http://dx.doi.org/10.12665/J123.Siegel.

Simonsen, Rasmus R. "A Queer Vegan Manifesto." *Journal for Critical Animal Studies* 10.3 (2012): 51–80, November 26, 2014. http://www.criticalanimal-studies.org/volume-10-issue-3-2012/.

Singer, Peter. *Animal Liberation: The Definite Classic of the Animal Movement.* New York: Harper Perennial, 2009.

Stanescu, Vasile. "'Green' Eggs and Ham?": The Myth of Sustainable Meat and the Danger of the Local." *Critical Theory and Animal Liberation*, edited by John Sanbonmatsu, 239–255. Plymouth, UK: Rowman and Littlefield, 2011.

Tally, Robert T. *Utopia in the Age of Globalization: Space, Representation, and the World-System.* Houndmills, Basingstoke, and Hampshire, UK: Palgrave Macmillan, 2013.

Tuomisto, Hanna L., and M. Joost Teixeira de Mattos. "Environmental Impacts of Cultured Meat Production." *Environmental Science & Technology* 45 (2011): 6117–6123.

Vattimo, Gianni. "Utopia, Counter-Utopia, Irony." In *Existential Utopia: New Perspectives on Utopian Thought*, edited by Patricia Vieira and Michael Marder. New York: Continuum, 2012.

Žižek, Slavoj. *The Sublime Object of Ideology.* London: Verso, 1989.

PART IV

Reflections

CHAPTER 8

Utopia and Biopolitics: The Need for an Ethics in Biotechnology

Cameron Barrows

Utopia and biopolitics have become integral to the sciences both in praxis and in theory. Since the Enlightenment, science has focused upon the improvement and categorization of the body but has seemed oblivious to its own utopian ideology. If scientism is the way of the future, then it is imperative that we understand the faults of it as an ideology and the way it approaches and perhaps distorts not just the physical body but also subjectivity.

The ethical dilemmas of modern-day science must be answered through an act of philosophy. This is of the utmost importance when we are discussing biotechnology and the role it plays in relation to our body. The practical methodology of modern science and biotechnology is to construct technological solutions to biological problems. In the blending of the bodily with the technological, we arrive at the state of a modified body, a modification that is essential to the development of a new ethics for understanding the body, biopolitics, and the Other in a world experiencing a paradigmatic shift in technology and its relation to the body. The body exists as an organic structure—a blending of nature and culture, human and nonhuman—and as a subjective entity that possesses consciousness and a mind that exists within the corporeal realm.

The body is both a physical and metaphysical object that has become objectified through the methodology of the modern sciences. There is no ethics if one sees the Other only as object. Ethics begins when we view the Other as subject. But ethics is not just the acknowledgment of the subjectivity of Other, rather it is the sustained act of how to treat and interact with the Other. This understanding of the Other is crucial to our understanding and critique of science and biotechnology.

If future progress is toward biotechnology and the construction of the human body into that of a modified subject, both machine and human, we must examine and radically doubt the empirical sciences if they are to become more incorporated into our Being, that is, our ontology. The empirical sciences are unable to examine themselves; for example, the science of biology is conducted through observation, a facet of perception. But biology as a science cannot study the act of perception; it cannot step outside of itself and examine the act of perception as a metaphysical act. This is the problem with the empirical sciences: they fail to understand the subjectivity of the subject and, because of this failing, the sciences cannot understand ethics. Ethics must be based upon the subjective experience of the Other. With the rise of biotechnology and the cult of science as the dominating order of the world and probably of our future, ethics should be the standard by which we determine our future actions toward other subjects.

In establishing an ethics concerned with technology and its relation to the body, we must first establish the politics of the body and the role of technology in this politics. A new ethics must first recognize that the vision of technology is not a vision of ethics but rather one of "progress." The search for technological solutions is one that—incorrectly—regards the body as an inherently flawed object that must be fixed. But in order to understand the politics of the body as it relates to the scientific epistemology, we must first undertake an examination of the history of the body.

The rise of Rationalism led to a new understanding of the body, both physically and metaphysically. The body became detached from nature and this schism allowed for the viewing of the body as an object rather than a subject. This disregard for the subjectivity of the body has allowed science to develop the methodologies of today. If we examine the protocol for human subject research and clinical trials, there arises a litany of ethical issues; not just ones that pertain to physical or emotional harm per se, but rather the dehumanization of the trial participant that comes from divorcing the subject from his or her subjectivity.

Modern science views the world and the human being within it as objects that can be bettered. If we examine the history of science, we can observe that many scientific discoveries and inventions have been lauded as great because they are viewed, ideologically, as "progressive."

This is not to suggest that scientific inventions such as vaccines, medicine, and certain technologies have not saved countless lives. But science does not publicly recognize its progressive nature, and it is this progressive nature within the ideology of science that must be accounted for. Without this acknowledgment, progress for the sake of progress has become the *modus operandi* of contemporary science.[1] This modality is inherently political, for scientific methodology and epistemology contains within it a utopian political ideology: that it will make the world and humanity better through scientific progress. This problem must be addressed explicitly if science is to further become a part of our ontology.

This utopian ideology begins within the scientific method itself. When an experiment is conducted, the subjects of the experiment (whether cells, animals, or humans) must become objects. This is where ethics begins to lose its connection with a care or concern for the Other. Even with the existence of such things as biomedical ethics and ethics that must be strictly followed especially in experiments involving humans, the objectification of a human subject devalues human life and experience, merely making it a statistic within a study. Through the process of experimentation, the scientist begins to embody the ideology of the science. The scientist-experiment relationship is one that is based upon the objectification of the Other, an objectification that inherently exhibits a lack of concern for the Other. The problem is not the lack of ethics within individual experiments, but the lack of an ethical awareness within the scientific methodology itself.

Since its inception in the Enlightenment, modern science has attempted to distance itself from the world of the real. It purports to provide an unbiased and objective view of the world and of humanity; this is simply not true. Objectivity cannot exist within a discipline in which the proponents of that objectivity are subjective themselves. The admittance of subjectivity into the scientific process would allow for the construction of a new and better ethics, especially if science and biotechnology become further engrained within our future ontology.

The ideology of science is often forgotten or ignored within the scientific process. Yet, ideology is inescapable; every field of study is rooted in ideology. In examining the scientific community's ideological roots, especially as it pertains to biotechnology, the ideas of Michel Foucault are most pertinent. Foucault's approach to ideology is centered around the discourses of power and knowledge as they relate to the subjugation of bodies and the advent of bio-power. Louis Althusser, Foucault's teacher and mentor, writes in his seminal essay *Ideology and Ideological*

State Apparatuses, "Ideology is a 'Representation' of the imaginary relationship of individuals to their real conditions of existence."[2] This thesis posits that the subject is only ideologically present in reference to the working conditions that surround him/her, a notion founded upon a central concept that ideology has no history of its own, but is instead a kind of historical distortion. Althusser writes, "Ideology has no history, for it is merely the pale, empty and inverted reflection of real history."[3] This conception of ideology as a distorting mirror of history offers confirmation for the idea of the history of science as progress for progress' sake, even though it is never revealed as such.

How do we resist the current ideology of the state of science and its objectification of the universe? The teleology of science is an impossibility: the viewer, the subject, must exist, but within a world where all falls under the objective panopticon of science, the subject—the viewer—does not exist. In order to resist such an absurd objectification, the subject must enter a state of suspension. This resistance is predicated upon an ethics. Yet within this ethics, one must resist the classification of existence that reduces the body to a mere label such as man, citizen, patient, and so on. This resistance is also futile, for we exist within the scope of science; it permeates the essence of our culture and therefore our ontology. The only true mode of resistance exists in the complete revolutionizing of the sciences in regard to both its epistemology and methodology.

With regard to the creation of a body politic in relation to ideology, the body of the political subject is constituted through the function of the state and is also constructed through the sovereign in the dialectic that occurs between the sovereign and the subject. Foucault writes in his essay *The Subject and Power*, "My objective, instead, has been to create a history of the different modes by which, in our culture, human beings are made subject."[4] For Foucault, the production of the subject seems to arise from the relations of power and knowledge and also the relation between the sovereign and the Other. This is seen in the creation of the scientific subject, the subject that is created by science, the hysterical woman, the insane patient, and the pervert. This mode of objectifying beings and making them into subjected objects is another of Foucault's means of producing the subjected. All of these modalities of production rely upon one notion though the creation of being as the Other: the political subject as Other in regard to the sovereign, the scientific subject as other to the scientist, and so forth. It is this mode of Othering that allows for the production of the subject in the modern state apparatus.

Foucault writes further in his lectures at the Collège de France in January and April of 1978 in a book titled *Security, Territory, Population*:

> By this I mean a number of phenomena that seem to me to be quite significant, namely, the set of mechanisms through which the basic biological features of the human species became the object of a political strategy, of a general strategy of power, or, in other words, how, starting from the 18th century, modern Western societies took on board the fundamental biological fact that human beings are a species. This is what I have called bio-power.[5]

This notion of bio-power is implicit within scientific praxis and methodology. It allows for control of the anatomo-politics of the human body, which is done through the regulation of populations and the control over the body of the individual through various means, whether medical, biological, or psychological experiments, or state medical practices. The body becomes not a place of the individual, but a space of the state; thus, it becomes a product of state egalitarianism in which all bodies are equal. This results in the body becoming a collective space in which the subject gets grouped into boxes, which in turn results in a situation that Foucault describes as "docile bodies." This notion of the docile body in anatomo-political realm allows for what Foucault describes in *Discipline and Punish*, "Discipline sometimes requires enclosure, the specification of a place heterogeneous to all other and closed in upon itself."[6] This notion of space or place in regard to both the docile body and the anatomo-politics requires us to see that ideology is not just a coercive tool used by the state to influence the consciousness of the people, but also that it has very real physical implications and makes its mark upon the body in physical manifestations.

Within scientific discourse, the body has become separated from its subjectivity. This change in the discourse of the body is one that can be traced back to the ideas of the Enlightenment, which have also influenced our thinking on ethics. Immanuel Kant, a preeminent Enlightenment thinker, proposed that ethics arises from reason and that, through our use of rationality and logic as so-called rational beings, we can arrive at what are considered traditional Western ethics (the ethics of the Judeo-Christian religious tradition). One proposed role of the body during the Enlightenment can be seen in the writings of the Marquis De Sade. Sade, in his philosophical discussions in *Philosophy in the Boudoir*, plays upon the notions of Enlightenment rationalism, especially its discourse about the body. For Sade, the body is a vessel, a material *tabula rasa*; this is grossly significant in his description and understanding of the body and its role within the societal ethics of the Sadean universe. Sade sees the human body as broken or fragmented; he suggests that we must fill the body and plug the holes and gaps. The body for Sade becomes a metaphor for the world as seen through the eyes of scientific rationalism, the predominant methodology and epistemology of the Enlightenment. And, the body has continued to be viewed and treated in such a manner ever since. Science and medicine still approach the body as a fragment, not one divorced from the subjectivity of the human being, but one that is inherently flawed.

This problem is rooted in the empirical sciences' view that perceives objects and events as purely objective entities. If we assume that all events and objects are objective, then we must deal with the overarching presence of the human subject as subjective and not merely as another objective entity that happens to have subjectivity. The problem with the empirical sciences on this matter is that the human being becomes merely an object of inquiry that happens to possess subjectivity. This empirical understanding of the subject is one that must change in order for the empirical sciences to grow and begin to understand the bizarre nature and interactions between the subject and the world around, which we deem the object. It is this reliance upon the objective as truth or the objective as accurate that seems to put the subjective in the realm of inconsistencies and the lack of rigor. But how can we have objective truth without the cognizance of the subjective human mind? Truth exists only through our ability to recognize it and comprehend it. Truth is merely a subjective notion and a manner of organizing the world around us into axioms and theorems to which a collective group of individuals adhere, for there is no inherency to any object or even to our ordering of the world. The world is merely chaos within which we, through our subjective mindset, recognize a pattern; we assume that recognition of a pattern of events is truth.

The problem regarding what constitutes truth or knowledge is that we can never approach any knowledge from an *a priori* or truly objective standpoint. This means that all knowledge can only be known through the lens of subjectivity. It is only through this consideration that we are able to see knowledge as it truly is. The reliance upon the objective nature of reality or the universe as a central axiom of the current empirical sciences is one that still considers the event and time as objective entities. Time is seen as an objective field of influence in which we exist, but it is equally likely that we posit time as the sensation of duration through our experience of movement or change. This notion that time is only an aspect of motion as perceived by the human intellect leads us to develop an epistemology of physics that is phenomenologically based and ultimately relies upon the subjective knowledge of the human subject. The idea that we can develop an understanding of the world that relies upon intuition rather than experimentation would be a major systematic and axiomatic change if we were to develop a new science, a new kind of scientific utopia best found in the practical methodology of the sciences.

Contemporary science exhibits utopian ideals in a number of ways. One of these is the insistence that science can answer problems empirically, even if these problems do not fall within the paradigm of empiricism. For example, the way in which science has tried to understand consciousness or morality seems reductionist at best and utterly asinine at worst. The issue with science trying to answer these problems through an empirical methodology is that it does not allow for an understanding of the full complexity of these issues. Science must adopt a dialectical approach, for the scientific paradigm is becoming increasingly narrow. The presupposition of certain axioms and hypotheses as objective truths has allowed us to construct the Borromean knot that is contemporary science.

Another problem that has arisen out of this muddled philosophical thinking and logic is our understanding of truth. For there are two types of truth: universal objective truth and local subjective truth. Universal objective truth is the one most of us recall when we define concepts such as truth, infallible, universal, or axiomatic. But the truths that we have come to rely on and that are common in our world are the subjective truths, those that are truth in regard to a certain situation, to a place we occupy, an ideology we hold.

Utopia exists within any ideology; the reliance upon hope and the betterment of the future is a very appealing thought for human beings. The utopian nature of science in both its methodology and epistemology is one where the world can easily be ordered according to certain axioms and the postulates that follow the said axioms. But the world is infinitely more complex than that, and there are certain objects and interactions that cannot be reduced and explained by a set of axioms. One of these is ethics. The concern and care for the Other are what allow for the production of ethics; this ethics only exists within systems that see another human being as a subject rather than an object. Ethics is not contrary to truth. It does not confuse the relations to the phenomena, for it is pre-philosophical in that it must be established before we construct a common ontology. Ethics is what allows for a movement toward truth. We need the relationship between subjects in order to discern truth in the world. Ethics does not undermine the questioning of the self, but rather it extends the questioning of the self through the questioning of the Other. An ontology is either incomplete or inconsistent if it cannot take into account the Other.

The ideology of science has lead to countless deaths and atrocities. We dismiss these horrific acts (e.g., the Holocaust) and say "never again." Yet we still adhere to the Enlightenment ideals that led to those acts, despite the massive damage and cost of human life it has accrued and the chance that these acts will be committed again. Modern science has moved beyond the ideals of its early, Enlightenment beginnings. The use of science now has become a weapon to be wielded against objects, and it allows for ordering, which is only partial understanding. Science has become a game of power and knowledge; it has been used as a means to exploit the poor and uneducated people of the world, which calls into question the very use and ideology of science itself. Can science exist within an ethical paradigm? If it cannot, this must lead us to question the ethical validity of science in the first place. Science exists as both methodology and worldview, which relies heavily upon the ideals of logic and empirical thinking. But can logic and empiricism exist within an ethical framework? And if the above is not true, that science cannot coexist within the world of ethics, what does this tell us about the state of science?

The state of science must change in order for science to continue to ask questions and seek answers. If science continues in the trajectory that it has set at the moment, more and more questions will arise that science will be unable to answer due to its narrow and specialized nature. Questions regarding ethics or our very ontology will remain unanswerable. However, perhaps it is best that they remain unanswerable within a scientific context, for this would allow science to develop a dialectical relationship with philosophy. This relationship would allow philosophy to critique science on a macro scale, whereas science could inform philosophy by further understanding the fundamental elements and systems that make up our world. The attainment of a certain symbiosis between science and philosophy would allow for a more complete and informed science and philosophy. It would allow us to move our thoughts into the world in a rigorous fashion. But this means ethics and the acceptance of subjectivity must become the new foundation for the science of the future.

With the increase in the role that science plays in relation to our lives and bodies, the ethical must become well-grounded within it. The ethical demand of science is not one that will tell us what action to take, but rather that we must act. The subject actor is where ethics begins; it is the subject's relation to the other, and the gap between them, that is the space where ethics must take place. Ethics is not certain and it never can be. Ethics is inherently an impossibility; it cannot take into account in its inception the chaos of the world. But this does not mean that we abandon ethics; rather, we must situate ethics in an essentially ontological manner. Ethics must become an inherent aspect of the subject's being, for ethics must become established through the subject's realization of a radical transcendental alterity. It is in this way that ethics becomes part of the nature of ontology.

Notes

1. This is not to say that scientific inventions such as vaccines, medicine, and certain technology have not saved countless lives; however, we must begin to account for the progressive nature that is denied within the ideology of science.
2. Vincent B. Leitch, et al. (eds.), *The Norton Anthology of Theory & Criticism*, 2nd ed. (New York: W. W. Norton, 2010), 1350.
3. Leitch, et al., *The Norton Anthology of Theory & Criticism*, 1349
4. Hubert L. Dreyfus, and Paul Rabinow, *Michel Foucault: Beyond Structuralism and Hermenuetics* (Chicago: University of Chicago Press, 1983), 208.
5. Michel Foucault, *Security, Territory, Population Lectures at the Collège de France 1977–78* (New York: Palgrave Macmillan, 2007), 1.
6. Michel Foucault, *Discipline and Punish: The Birth of the Prison* (New York: Vintage Books, 1995), 141.

Bibliography

Dreyfus, Hubert L., and Paul Rabinow. *Michel Foucault: Beyond Structuralism and Hermenuetics*. Chicago: University of Chicago Press, 1983.

Foucault, Michel. *Discipline and Punish: The Birth of the Prison*. New York: Vintage Books, 1995.

———. *Security, Territory, Population Lectures at the Collège de France 1977–78*. New York: Palgrave Macmillan, 2007.

Hitchens, Christopher. *The Missionary Position: Mother Teresa in Theory and Practice* (New York: Verso, 1995).

Leitch, Vincent B., William E. Cain, Laurie Finke, Barbara Johnson, John McGowan, T. Denean Sharpley-Whiting, and Jeffrey J. Williams (Eds.). *The Norton Anthology of Theory & Criticism*, 2nd ed. New York: W. W. Norton, 2010.

Wise, Jacqui. "Pfizer Accused of Testing New Drug Without Ethical Approval." *BMJ* 322 (7280): 194.

Contributors

Cameron Barrows is a native of Wenatchee, Washington. He studied philosophy and literature at Bard College at Simons Rock. He is currently a graduate student at St. John's College in Santa Fe, New Mexico. His work focuses on ontology, ethics, language, and science.

Andrew Byers is a cultural and social historian of the twentieth-century United States working as a visiting assistant professor of history at Duke University. His research interests focus on the intersections of law, science, sexuality, the body, and the military. He is currently examining the US military in the early decades of the twentieth century within the context of changing legal and psychological frameworks as well as social and moral reform debates. His dissertation, titled "The Sexual Economy of War: Regulation of Sexuality and the US Army, 1898–1940," explores how the US Army of the early twentieth century sought to regulate and shape the sexual cultures, practices, and behaviors of soldiers and the civilians with whom they came into contact around the world.

Elena L. Cohen is an activist, attorney, and doctoral candidate. Having graduated from Benjamin N. Cardozo School of Law, Ms. Cohen is now completing her PhD in political theory at the Graduate Center of the City University of New York. She is the current president of the New York City Chapter of the National Lawyers Guild.

Arpita Das currently works with a regional women's rights organization focusing on sexual and reproductive health and rights issues in the Asia and Pacific region; she holds an MA in Social Work from Tata Institute of Social Sciences, Mumbai, India and an MA in Women's and Gender Studies from the University of Łódź , Poland and Central European University, Hungary. Her professional experience includes working and writing on issues of gender-based violence, sexual and reproductive health and rights, sex selection, sexuality education, and reproductive technologies. She serves as co-chief editor of the Graduate Journal of Social Science, an open-access, peer-reviewed, multidisciplinary academic journal. Her academic and research interests include gender, gender-based violence, sexuality, intersex issues, disability and sexuality, reproductive technologies, and biopolitics. She can be reached at arpitadas07@gmail.com.

Evie Kendal is a feminist bioethicist and literary critic from Melbourne, Australia. Her research interests include representations of biotechnology in science fiction, legal and ethical issues for end-of-life care, and feminist issues in young adult literature and films. She is currently affiliated with the Centre for Human Bioethics and School of Languages, Literatures, Cultures and Linguistics at Monash University, researching representations of ectogenesis and other reproductive biotechnologies in science fiction and popularc ulture.

Selena Middleton is a doctoral student with the Department of English and Cultural Studies at McMaster University in Hamilton, Canada. Her dissertation is titled "The Lands between Pasture and Apocalypse: An Ecocritical Approach to Wilderness in Contemporary Science Fiction." The project focuses on the pastoral and apocalyptic modes, and wilderness as a liminal space between the Edenic beginnings of origin mythology and dystopian futures in environmentally themed science fiction. Her work on wilderness in Quaker literature was published in the Winter-Spring 2013 issue of *Quaker Theology*. Middleton is also currently in the final stages of editing her own science fiction novel, *Mouthful of Fire*.

Rasmus R. Simonsen is currently teaching in the Centre for American Studies and the Department of English at Western University, Canada. His primary research focuses on the relationship between rhetoric and sexuality in nineteenth-century American literature. In 2012, he published "A Queer Vegan Manifesto" in the *Journal for Critical Animal Studies*. It is the first essay to capture the personal and theoretical aspects of going vegan from a queer perspective. Since its initial publication, it has been published in book form by Ortica Editrice in Italian as *Manifesto Queer Vegan*.

Patricia Stapleton is a comparative political science and public policy scholar. She currently teaches at Worcester Polytechnic Institute in Massachusetts, where she is also the director of the Society, Technology, and Policy Program. Her research interests cover the regulation of biotechnology, in both food production and reproductive medicine. Stapleton's dissertation explored the differences between the European Union and France in regulatory policy making for genetically modified organisms destined for human consumption. She is currently researching the possible impacts that the Affordable Care Act might have on reproductive health services in the United States.

Index

able-bodiedness, 16, 30, 32, 43, 51
abnormal, 42, 43, 44, 45–7
Agranoff, David, 167–8, 178, 180
Althusser, Louis, 195–6
American Association for Health, Physical Education, and Recreation (AAHPER), 16, 21
artificial human reproduction, 94
assisted reproductive technologies (ART), 63, 65–70
    regulation of, 64, 70–4, 76–7

Baartman, Saartje, 45–6
Bacigalupi, Paolo, 122, 127
Bieber Lake, Christina, 120–1, 134–5
bioethics, 90, 91, 92, 94, 96, 97, 98, 99, 100, 101, 102, 103, 104, 105, 106, 107, 108
biopolitics, 1, 2, 3, 13, 15, 32, 42, 44, 50, 90, 91, 92, 97, 99, 102, 103, 104, 108, 168, 170–1, 193
    biopolitical intervention by the state, 1, 4, 42, 51, 170
    spatial dimension of, 146–9
biopower, 1, 5, 143, 144, 170, 197
biotechnology, 94, 95, 96, 97, 98, 99, 102, 103, 104, 105, 106, 107, 108, 119, 121, 123, 124–5, 126, 128, 133–4, 193, 194, 195
    reproductive, 49, 67, 74, 91, 93, 100
body/bodies, 1, 2, 3, 13, 14, 15, 24, 26, 28, 30, 32, 41, 42, 46–7, 51, 65, 73, 92, 123, 127, 131, 132, 133, 193, 194, 196, 197, 198
    docile, 14, 144, 197
    monstrous, 46
    regulation of, 1, 2, 3, 4, 5, 13, 14, 15, 92, 197
    softness, 23
branding, 144, 147
Brave New World, 90–2, 94–5, 96, 97–9, 105–6
business, 145–6

Canning, Kathleen, 3
carnivorism, 179–80
Catts, Oron, 169, 172–6
Centers for Disease Control and Prevention (CDC), 65–6, 71, 72
children, 15, 19
citizenship, 1, 3, 5, 13, 15, 20, 30–1
civil defense, 26
cloning, 93–4, 95, 96, 98, 103, 105, 106
Cold War, 22–8
colonialism, 121, 123, 124, 125, 127, 128–9, 130, 131, 133, 135

De Sade, Marquis, 197
Deleuze, Gilles, 2
deviation, 150, 152, 153
disability, 41, 42, 43, 44, 45–6, 47–8, 51, 53–4, 69, 75
discipline, 2, 143, 144–6, 197
Dolly, 95, 96
draft, 16, 28
    rejection, 17, 18, 28
dystopia, 78, 89, 91, 94, 98, 99, 101, 105, 106, 107

ectogenesis, 105, 106, 107, 108
Eisenhower administration, 23
Enlightenment, 193, 195, 197, 198, 200
ethical consumerism, 182–3
ethics, 74–5, 121, 126, 128, 135, 176–7, 180, 193, 194, 195, 196, 197, 198, 199–201
eugenics, 43, 47–51, 100, 101
eutopia, 89, 107
eviction, 146, 155
exceptions, 42, 43, 44, 51–4

First Amendment, 143, 146
fitness, 18, 22, 24
Foetal Bovine Serum (FBS), 169–72, 177, 180, 184
food security, 121, 125
Foucault, Michel, 1, 2, 3, 14, 42, 51, 92, 141–6, 149–52, 153–5, 170, 181–2, 195, 196–7
　hegemony, 142, 151
　heterochronies, 151
　heterogeneity, 150
　inversion, 152
　milieu, 146
　security, 146
　sovereignty, 144
Frankenstein, 90–2, 96, 99–100, 102, 119–20
freedom, 152

Gattaca, 90–2, 100–1, 104
genetic discrimination, 74–6, 91, 100, 104
genetic engineering, 43, 47–51, 91, 98, 99, 104
genetic enhancement, 92, 101, 104
genetic testing, 52, 64, 67, 72, 78
governmentality, 3, 46, 51
Green, Ronald M., 97, 98
Griffith, Nicola, 126–7
Grosz, Elizabeth, 3, 14

happiness, 181–2
　futurity, 182

heterotopia, 149–52, 153–5
homogeneity, 150
hubris, 91, 99, 102
Huxley, Aldous, 90–1, 92, 93, 94, 95, 97, 98, 101, 106, 107
hygiene, 145–6

ideology, 193, 195, 196, 197, 199, 200
in vitro fertilization (IVF), 67, 69, 75–6, 94, 104, 105, 107
in vitro meat, 168–70, 172–7, 178–9, 180, 184
India, 47, 49
infertility, 69, 75
intellectual property rights, 123, 124, 127
intersex, 41, 42, 43, 44, 46–7, 53–4

Jameson, Frederic, 107
Johnson administration, 27–8

Kant, Immanuel, 197
Kass, Leon, 97, 98
Kennedy administration, 27
Kress, Nancy, 127

Latour, Bruno, 119–20, 122, 135
Levitas, Ruth, 4, 64, 65

militarization, 30
Monsanto, 124, 129
More, Thomas, 176

narrative, 121, 126–7
nation, 15
national security, 13, 15, 29–30
New York Police Department (NYPD), 144
Niccol, Andrew, 90–1, 101
Nineteen Eighty-Four, 94
normalization, 41, 42, 45, 46, 53–4

Obama administration, 29
Occupy Wall Street, 141, 143–6, 147–9, 152–5

ontology, 194, 195, 196, 199, 200, 201
Orwell, George, 94
Other, 193, 194, 195, 196, 199, 200

paid surrogacy, 70–1
patriotism, 22
Pechlaner, Gabriela, 125
People for the Ethical Treatment of
    Animals (PETA), 168–9, 178, 180
Pistorius, Oscar, 43, 51–2
playing God, 91, 99, 100
population, 1
post-animality, 178–9
posthuman, 103, 106, 107, 179
power, 3, 14, 45, 91–2, 100, 101, 108,
    122, 124, 128
pregnancy, 69
preimplantation genetic diagnosis
    (PGD), 67–9, 74, 75, 76, 77
Presidential Physical Fitness
    Award, 28
President's Council on Physical
    Fitness, 27
President's Council on Youth
    Fitness, 23
protest, 152
Pump Six and Other Stories,
    127–8, 133–4

rationalism, 194
representation, 152
reproduction, 42, 47–8, 50, 51, 70,
    97, 99, 101, 106, 107, 108, 124,
    125, 133
reproductive health, 65
resistance, 2, 14, 29, 32, 92, 101, 125–6,
    127, 128, 129, 133, 149, 152, 153
Roosevelt administration, 18, 20

Sargent, Lyman Tower, 89, 108
science, 193, 194, 195, 196, 198,
    199, 200
science fiction (sf), 89, 90, 91, 92, 94,
    95, 96, 98, 99, 100, 102, 103, 104,
    105, 107, 108, 109

scientific imaginary, 90, 96
Selective Service, 16, 17, 20
sex selection, 67, 75
Shelley, Mary, 90–1, 94, 99, 100,
    102, 119
Shiva, Vandana, 121, 123, 124, 125–6,
    127, 128, 132
Shukin, Nicole, 170–1
Silver, Lee M., 95–6, 101, 104
space, 141, 142–6, 149–50, 152
    resistance, 149–52
spatial tactics, 144, 146, 149
speciesism, 170
sports, 43, 51, 58
Squier, Susan M., 103, 108
subject, 194, 195, 196, 197, 198,
    199, 201
subjectivity, 193, 194, 197, 198, 200
Suleman, Nadya, 73–4
surveillance, 2, 144, 146, 147, 149
Suvin, Darko, 89

technologies, 42, 43, 44, 49–50, 182
    reproductive, 50, 63, 66–70, 73
technophobia, 90, 95
technoscience, 92, 108
terminator seed, 124–5
thought experiment, 89, 95, 105
total war, 20–1
Turner, Bryan, 2, 65

utopia, 2, 3, 4, 5, 15, 22, 29, 32, 54,
    64, 73, 89, 90, 92, 95, 96, 98, 99,
    102, 103, 104, 105, 106, 107, 108,
    122–3, 134, 168, 172, 175–6, 177,
    184, 193, 195, 199
    scientific utopia, 199
    utopian aspirations of the
        state, 3, 4
    utopian drive, 63, 70
    utopian fiction, 122
    utopian veganism, 173, 183

veganism, 167–8, 173, 176, 177, 180,
    181–4, 185

Williams, Raymond, 122
The Windup Girl,
  127–34
World War I, 17
World War II, 16–22, 47

Žižek, Slavoj, 177
zombies, 167–8
zoopolitics, 170
Zuccotti Park, 141, 144, 146, 152, 155
Zurr, Ionat, 169, 172–6

The manufacturer's authorised representative in the EU is Springer Nature Customer Service Centre GmbH, Europaplatz 3, 69115 Heidelberg, Germany. If you have any concerns regarding our products, please contact ProductSafety@springernature.com

Printed and bound by CPI Group (UK) Ltd, Croydon, CR0 4YY
23/03/2026
02076449-0015